THE HIGH COST OF HIGH TECH

A
CORNELIA
AND
MICHAEL
BESSIE
BOOK

■ ■ ■ ■ ■

HARPER
& ROW,
PUBLISHERS
New York
Cambridge
Philadelphia
San Francisco
London
Mexico City
São Paulo
Singapore
Sydney

The Dark Side of the Chip

THE HIGH COST OF HIGH TECH

BY
LENNY SIEGEL
AND
JOHN MARKOFF

THE HIGH COST OF HIGH TECH.

For information address Harper
& Row, Publishers, Inc., 10 East
53rd Street, New York, N.Y.
10022. Published simultaneously
in Canada by Fitzhenry &
Whiteside Limited, Toronto.

FIRST EDITION

Designer: Elissa Ichiyasu

This book is set in 10-point
ComCom Vermillion. It was
composed by ComCom
and printed and bound by
The Haddon Craftsmen.

Library of Congress Cataloging in
Publication Data
Siegel, Lenny.
The high cost of high tech.
"A Cornelia and Michael Bessie book."
Includes index.
1. Computers and civilization.
I. Markoff, John. II. Title.
QA76.9.C66S537 1985 303.4'834 84-48622
ISBN 0-06-039045-X

---

85 86 87 88 HC 10 9 8 7 6 5 4 3 2 1

Contents

■ ■ ■ ■ ■

■ ■ ■ ■ ■

ACKNOWLEDGMENTS

We would like to extend our appreciation to our parents, and Lenny Siegel particularly wishes to thank his wife, Jan Rivers, and her family for their support and patience during the creation of this book. In addition, we thank Walter Cohen, Randy Schutt, Deirdre Kelly, and other friends, associates, and supporters of the Pacific Studies Center for their help.

Chapter One
COMPUTER CITIZENSHIP: THE PERIL AND THE PROMISE

■ ■ ■ ■ ■

Several years ago, at a military trade show held in a glittering Las Vegas hotel, an earnest Air Force colonel tried to describe what electronic warfare meant to him. Searching for an appropriately upbeat note on which to end the interview, the colonel, who earlier in the day had led a press tour of a military electronics exhibition, exuded the confidence of a true believer.

"Electronic warfare is more than just a science," he argued. "It's an art.

"More than that," he implored, groping for precisely the right word, "it's . . . it's a religion."

The image of this colonel, a colorless figure in uniform, is haunting. Here is a man who is destined to fight the next war from a sterile, air-conditioned command bunker hundreds or even thousands of miles from the actual battle zone, and yet he speaks with the zeal of a Roman gladiator. He bitterly hates an abstract enemy he will never see, except perhaps as a ghostly signature on a video display screen. Yet he is clearly ready for battle.

This captivation with technology is not confined to the military. A few years ago, Steve Wozniak, the designer of the Apple II Computer, struck a fusion between technology and youth culture by sponsoring the elaborate "US" festival in Southern California to celebrate both rock music and personal computers. Wozniak and his colleagues at Apple Computer blend a faith in the power of the microprocessor with personal values based on social concern. "Woz," as he is known, became a legendary entrepreneur in Silicon Valley almost in spite of himself. In a few short years he catapulted from the presidency of the Electronics Club at a Sunnyvale high school to become one of the world's richest men. However, he quickly tired of management at Apple and quit the company, returning later, by choice, as an engineer, only to strike out on his own again.

More recently, Stewart Brand, a guru of the counterculture and editor of the *Whole Earth Catalog,* tried to blend his enthusiasm for personal computers with a broader cultural movement by publishing the *Whole Earth Software Catalog.* Brand, who was known as a spiritual leader of the anti-technology, back-to-the-land movement that swept the country in the late sixties, now writes articles comparing the merits of different software packages. He says: "Computers rejuvenated our minds; we haven't learned so intensely since college. Computers reconnected us to the future."

Computer Citizenship: The Peril and the Promise

Throughout American society, microelectronics and the advent of the personal computer have engendered an excitement found almost nowhere else in our lives. What is so special about this technology, that makes possible the shared vision of the general, the engineer, and the philosopher?

It's simply that the microprocessor, a spider-like network of microscopic circuits etched on a flake of silicon less than half the size of your fingertip, has reversed a centuries-old trend by personalizing technology. It has rendered cars, telephones, ovens, and countless other devices more responsive to the human touch. The personal computer, a machine made possible by the microprocessor, is not only approachable by millions of people who previously felt trapped by faceless technologies and impersonal institutions; it is one of just a few things in many people's lives that they can control.

Popular in its familiar forms, microelectronics technology is increasingly seen as a panacea for America's political, social, economic, environmental, and military ills. Rising Democratic and Republican politicians promise to create a better investment climate for high-tech industry. Until the company's misfortunes mounted, it was chic to be labeled an "Atari Democrat."

Pop sociologists and futurists have decided that the impact of microelectronics is epochal and are declaring industrial America of the 1980s an "information society." They even go so far as to pick dates to mark the year of transition from the industrial to the information world. One writer suggests that the onset of the information age came in 1973, during the Arab oil embargo, when it became clear that we could no longer trust America's industrial infrastructure. Others are bolder.

John Naisbitt, author of *Megatrends,* suggests that the information era descended on us in the inauspicious year of 1956: this was the year that white-collar workers outnumbered their blue-collar counterparts for the first time.

The common conception of an information society is naively self-congratulatory. Its roots lie in the work of a group of American social thinkers led by sociologist Daniel Bell, who first proposed the concept of a "post-industrial" society. For them, the end of the industrial era was cause for celebration. They were caught up with exuberance in the belief that the major social problems of post–World War II America were soluble with the judicious application of technology and rational planning.

The turbulent sixties demolished Bell's myth, but claims are being made once again that there are technological quick fixes for deeply rooted social problems. We are indeed living in a world increasingly dominated by the collection, processing, and distribution of information. But the advent of the silicon age has merely altered, not eliminated, age-old problems.

Today, the leading prophets of high tech are no longer social scientists; they are entrepreneurial engineers. These men have performed their own modern alchemy, turning common sand—silicon dioxide—into reliable and inexpensive machines capable of manipulating numbers, words, and pictures far faster than the human mind.

The high priests of high tech, whose badges of success bulge from their pocketbooks, have inspired a blind optimism with their predictions of a computerized utopia. And like earlier generations of entrepreneurs before them, they equate social value with personal or corporate profit. "Don't tamper with our freedom to make money," they argue, "and technology will lift our collective burden, exercise our minds, and exorcise our conflicts."

■ The new Mecca for the high priests of high tech is Santa Clara County, California. In the early 1970s, insiders began to call the area Silicon Valley because of its high concentration of semiconductor manufacturers. Today, Silicon Valley is known universally, not only for the world's greatest gathering of microelectronics research and production, but as the symbol of the oft-repeated promise of prosperity evoked by high technology.

In the space of one month during 1984, Silicon Valley hosted delegations of economic development officials headed by the king of Sweden, the president of Austria, and the president of France. The queen of England had already led a delegation to the Valley the previous year.

During these visits, the role of chiefs of state is to underscore the commitment that their governments have made to high technology. They make the pilgrimage to Silicon Valley in an attempt to grab a piece of the chip—that is, to entice job-producing investment from the Valley's rapidly expanding high-tech corporations and to style their governments as forward-looking and progressive.

Such royal and presidential visits are carefully orchestrated to show off the best that high technology has to offer. Journalists and officials return to their homelands telling of a land where factories look like college campuses, unemployment is low, the tax base is solid, and life is comfortable.

Silicon Valley is populated by some of the best scientific minds of a generation. It is touted as a model for development both in advanced industrial society and in the developing world. For the information-rich—engineers, programmers, scientists, and managers—Silicon Valley is indeed a success story. Their knowledge and skills are in high demand, so they can afford to live in Santa Clara County's comfortable, affluent suburbs. Nearly all of these professionals hold college degrees; a majority have at least two. Most are white men.

But distinguished visitors rarely see the Valley's second society: the other half of the workforce, production and service workers who labor at the same companies as the professionals, but who otherwise are miles apart in pay, culture, and residence. Despite its information-society-of-the-future image, Silicon Valley still has industrial sweatshops and even illegal cottage industries, in which immigrant women assemble printed circuit boards in their homes.

The problems go much deeper. Portions of the Valley's water supply have been poisoned by leaks from high-tech industry's chemical tanks. Despite the vigorous enforcement of environmental regulations, numerous firms routinely pour toxic wastes down the drain. The San Jose Unified School District, where many of the women and minorities who make up the production workforce live, went bankrupt in 1983. Housing costs, driven up by the influx of well-paid professionals, are among the highest in the country. In fact, experts in and out of industry can't figure out how production workers manage the Valley's unusually high cost of living on their meager wages.

Still, even the lowliest Silicon Valley worker earns ten times as much as the women who package semiconductor chips in East Asia. Ironically, the most complex machines ever built are assembled by some of the lowest-income industrial workers in the world. Though high-tech executives warn that the Japanese are taking over, they cooperate with their Japanese counterparts while manufacturing their own components and devices in other parts of Asia.

The world of high-technology production is mirrored by a growing schism throughout society between the information-rich and the information-poor. Soon, it appears, computers and other microprocessor-controlled equipment will be firmly entrenched in virtually every workplace. But on the job, only a small information elite will really control the high-tech tools. Most workers, such as data entry clerks, who sit parked all day in front of their terminals, will be controlled or monitored by their machines. Furthermore, unless society finds new ways to guarantee employment or income, millions of workers displaced by the rise of high tech will be out of work and out of luck. And in the long run, even professional workers may be replaced by "intelligent" machines that won't think in the human sense, yet will still touch off a new round of workplace automation.

The "electronic cottage" envisioned by futurists promises a range of computer-based services for the information haves, from electronic mail to home banking. But the have-nots may find themselves stranded in their homes with no phones, decaying public libraries, and declining postal services. Like the car-less urban poor in America's central cities today, they will be locked out of the benefits of a remarkable new technology.

If Silicon Valley represents the promise of a technological utopia, it also epitomizes the peril of an Orwellian world out of control. Slowly but surely the average American has unknowingly accepted the electronic mapping of his or her entire existence. Potential tyrants have been honing their high-tech surveillance techniques, awaiting the opportunity to stifle dissent and crush nonconformity.

Microelectronics technology is the cornerstone of a new generation of weapons that endangers the entire human species. The most advanced U.S. weapons, whether intended for nuclear war or for conventional battlefields, contain smaller warheads and explosive charges than before, but they are much more deadly because they can be sped to their targets with incredible accuracy. Pentagon programs to imbue machines with "intelligence" may soon give autonomous, computerized satellites the ability to launch World War III. In the form of a little known but dangerous game of electronic combat, World War III is already here.

■ From the start, the development of microelectronics and computer technology was deeply intertwined with the requirements of the U.S. armed forces. Beginning with the use of early computational machines to build the first atomic bombs during World War II, to the design of missile guidance systems in the 1960s, the military defined America's technological priorities.

Although civilians invented the germanium transistor, the first semiconductor, at AT&T's Bell Laboratories in the late 1940s, Bell's work grew directly from government-sponsored solid state physics research conducted during World War II. Without the military's early investment in high-cost, low-volume components, it might have taken years before manufacturing techniques improved enough to put semiconductors into mass production. During the first five years that the transistor was available commercially, the Pentagon, particularly the Army Signal Corps, ordered millions of dollars worth of the new devices. At the same time, if the military had clamped a lid of secrecy on the infant semiconductor technology, the new industry would have collapsed.

From 1959 to 1961, research teams at both Fairchild Semiconductor in Silicon Valley and Texas Instruments independently developed solid state components that combined more than one transistor into "integrated circuits," or chips. In the early 1960s the Air Force, which needed to squeeze an intricate guidance system into the nose cone of the Minuteman intercontinental ballistic missile, subsidized the early development of chip technology with hundreds of millions of dollars in purchases. It is ironic that the technology that brought computers, video games, and electronic auto ignitions to the American consumer arrived in the deadly tip of a rocket designed to rain thermonuclear fire half a world away.

In the 1970s, the civilian economy gained control over the direction of microelectronics and computer technology from the Pentagon. Researchers at Intel—at the time a small Silicon Valley firm with no military contracts—invented the first microprocessor in 1971 at the request of a Japanese calculator manufacturer. A few years later, high-tech hobbyists in Silicon Valley started designing their own microcomputers, computers based upon a single microprocessor. One such hobbyist was Steve Wozniak, whose Apple II turned the data-processing industry upside down.

The Pentagon, still dependent upon chips, computers, and other advanced electronics technologies for its weapons, hasn't given up on the industry. In fact, it has launched a concerted research and development program to regain its position of control over microelectronics. And while spending hundreds of millions of dollars on microelectronics and computer science research, it is attempting to impose new controls on the export of high-tech goods and the public dissemination of advanced electronics research.

■ Not long ago, the city of San Jose outbid two other Silicon Valley cities to attract a high-technology museum, proposed to feature "hands-on" exhibits acquainting visitors with the wonders of microelectronics and computers. The excitement stimulated by the project, long before the first cornerstone has been laid, illustrates America's fascination with high tech.

Without even knowing what a semiconductor chip looks like, people know that its use is already making profound changes in the way they learn, work, and live. Of course, they want to learn more about the technology. How does it work? What can it do? How can they prepare themselves and their children for a world shaped by advanced electronics?

Most Americans don't realize that they have the power to shape the way that high technology is used. In fact, they—as voters, workers, consumers, and parents—may have as much influence over the future of high technology as high tech has over their future. *How* they learn about high tech, and *what* they learn, therefore, is a key to the future of our society.

The self-appointed heralds of the silicon age, futurists and high-tech executives, say that "computer literacy" is becoming an economic necessity. Everyone, they claim, needs to know how to program a personal computer, or at least how to use one. Those who remain computer-illiterate will soon find themselves frying burgers, digging ditches, tending furnaces, or on relief.

The proponents of computer literacy are correct to suggest that most of the American workforce may soon work with computerized equipment. Already the video terminal, the most recognizable emblem of high tech, is found almost everywhere, from the desks of stockbrokers to the sales counters of lumber yards. Ironically, even some of the least educated, lowest-paid workers in America—kids who serve fast food, for instance—already work with computers.

To operate most of that equipment, however, requires little more skill than it takes to operate an automatic teller machine at a bank. Computer manufacturers are investing hundreds of millions of dollars each year in placing more "intelligence" in their machines and making them simpler to operate. The knowledge necessary to run a typical microcomputer today may prove useless in five or ten years, just as the essentials of computer literacy, as taught only a decade ago, are obsolete today.

What the enthusiasts often forget is that no education is more important, if one is to succeed in the rapidly changing world of high tech, than a solid background in plain old literacy. Until someone devises a machine that can read our minds, our ability to communicate, either with machines or with people, depends upon our ability to read, write, and speak coherently. Language is the fundamental tool. And it is far more flexible and universal than the most remarkable of high-tech tools.

Mathematics, the other traditional basic skill, remains vital for adjusting to the world of high tech. Although with the help of simple high-tech devices, such as pocket calculators, one can survive without the command of arithmetic, the ability to frame problems is still essential. If any one ability stands out as a measure of an individual's likelihood of business or scholastic success, it is the ability to think logically and analytically.

If Americans merely prepare for personal success, however, they will have abandoned their rights and obligations as citizens. To fulfill their responsibilities and protect their collective interests, they must learn about the peril as well as the promise of high technology. They need a background not in computer literacy, as defined by the electronics industry, but in computer citizenship. Communication and reasoning skills, the prerequisites for personal achievement, are also essential to good citizenship in the computer age; but much more is required.

Computer citizenship means knowing enough about the social, political, environmental, and military implications of computer technology to make personal and public choices. Even if we never learn how to "boot a disk," or haven't even heard that terminology, we must learn to be good citizens in the information age. In fact, the vast resources now dedicated to teaching computer use may simply be diverting us from a more important learning process.

Americans need to be able to evaluate the threat that computerized record systems pose to individual privacy and freedom. Our communities must know the environmental risks of chip and computer production. We must understand that advances in electronics, not larger nuclear warheads, threaten to plunge us into global war. We must ask how microprocessors can be used to enhance, not dehumanize, the work of the typical factory, office, or service worker.

Today, the control of high technology lies in the hands of a relatively small number of experts, bureaucrats, and business executives. Without a widespread public understanding of the actual and potential costs of high tech, they will continue to make critical decisions about our collective future. If one believes, like Voltaire's Dr. Pangloss, that we live in the best of all possible worlds, then there is no cause for alarm. However, if we are to extend our democratic traditions and hopes into the information era, then universal computer citizenship is essential. Should we fail, we all will surely pay a high price.

The social costs of the continuing spread and evolution of high technology are great, but they are by no means inevitable. The challenge of microelectronics is to develop new technologies in a manner which minimizes their costs. High-tech industry doesn't have to pollute; jobs do not have to be boring; and our scientists and engineers don't need to work in an economy where the main outlet for their creative talents is the weapons industry.

Many who earn their living or their fortune from high technology will condemn as anti-technology every attempt to question high-tech innovation, as if technology and industry were neutral. Nothing could be further from the truth. Our machines and our sciences, as well as their uses, have been shaped by our past.

Today, the image of high technology is as shiny as the surface of the wafers of polished silicon from which integrated circuits are made. If we look closely, however, we see a reflection of ourselves. The information society reflects the world from which it emerged, with all its hopes and successes, problems, failures, and fears.

For society to reap the highest benefit from information technology, we must look beneath the surface to its dark side. We don't need to understand the physics of a microprocessor, the algorithms of a computer program, or the tax treatment of venture capital to make intelligent choices about America's high-tech future. But information about the *impact* of computers and microelectronics must be made available and comprehensible to all of us if we are to function as full citizens in the age of high technology.

Chapter Two
THE ALL-ELECTRONIC ARMS RACE

■ ■ ■ ■ ■

In the heart of Silicon Valley, at the crossroads of California Highway 237 and U.S. 101, looms a five-story-high windowless building known to many as the Blue Cube. A man who once worked there says the inside looks like a scene from a science fiction thriller: "Not only are there locks on all the doors, but there are locks on doors inside rooms with locks on doors inside rooms . . ."

The sign outside the high barbed-wire fence that surrounds the installation informs passersby that this is the Sunnyvale Air Force Station, but there are no bombers or jet fighters in sight. Instead, a cluster of giant parabolic antennae, like a jumble of topsy-turvy teacups, points skyward. No nuclear warheads are sheltered on its grounds. The heaviest ammo sits in the belts of the Station's security guards. Yet if the Soviets ever launch their nuclear-tipped missiles at the United States, Sunnyvale's modest little Air Force Station will be one of their first targets.

The Blue Cube is one of a handful of high-tech nerve centers linking together America's far-flung military forces. Without it, the American military satellite system would soon be worth little more than space garbage, even if every single spacecraft remained intact. Sunnyvale's men and computers monitor and control the orbits of at least fifty military satellites, and the Cube constantly receives and analyzes a stream of photographic and electronic data from America's orbiting sentinels.

The Cube's critical importance underscores a little known fact: microelectronics, not nuclear physics, is the key to making preparations for nuclear war. American nuclear scientists and engineers have refined the design and manufacture of nuclear weapons since they fired off the first H-bomb in 1952, but the weapons—horribly powerful, to be sure—remain essentially the same. The significant advances have come instead in command, communications, intelligence, guidance, and other systems based on silicon chips and computers.

Three decades ago, Soviet and American aircraft carried powerful nuclear weapons, but the world did not face the threat of instant holocaust. Before the development of microelectronics, there could be no reconnaissance satellites, communications relay spacecraft, or missile guidance systems, essential components of modern strategic warfare. In fact, during the early 1960s the Air Force ordered hundreds of millions of dollars worth of integrated circuits, before they were commercially available, to squeeze a complex guidance system into the nose cone of a Minuteman missile. As microelectronics made possible a new kind of nuclear war, the military's purchasing strength boosted the new technology to a point where it could have commercial applications.

■ Despite all the controversy surrounding President Reagan's Strategic Defense Initiative, or so-called Star Wars proposal, which would place explosive devices and beam weapons in space, the Pentagon has relied heavily upon orbiting electronic systems for more than two decades. Reconnaissance satellites have plotted the growth of the Soviet war machine; communications spacecraft have offered U.S. commanders instant contact with ships, planes, and ground stations everywhere. Even if the United States decides not to move beam or explosive weapons into space, space is likely to be an area of growing superpower contention.

Since spacecraft are less vulnerable to attack than fixed, exposed installations like the Sunnyvale Air Force Station, the United States and the USSR are moving vital missions into space as fast as technically possible. While constantly modernizing its orbiting communications and intelligence craft, the Pentagon is deploying the NAVSTAR satellite system, which will allow U.S. ships, submarines, and aircraft to fix their positions to some fifty feet in three dimensions. Research is now under way on a space-based radar system, designed to replace the country's international network of fixed air and missile defense radars, and the Pentagon is supporting research designed to automate and eventually move decision-making functions, such as those exercised by the Blue Cube, into orbit as well.

With both superpowers becoming dependent upon space to relay commands and target their missiles and bombs, it is not surprising that the United States and the

Soviet Union are developing space mines, missiles, and laser beam weapons designed to damage or destroy each other's satellites. War planners have come up with a neat rationale. "Space is a dandy arena, actually," claims one Pentagon scientist. "You've got to attract strategic war off the planet."

Such strategists forget, however, that space war is inseparable from war on the planet. The purpose of anti-satellite warfare is to ease the task of sending nuclear weapons to their terrestrial targets. Space is not a distant playing field where society can isolate death and destruction; it is the high ground of high-tech warfare.

Military planners want to seize that "high frontier" for ballistic missile defense, as well. Though few in the Pentagon believe that the United States will ever be able to field an anti-missile system capable of protecting the American population by fully neutralizing an onslaught of Soviet missiles, they see the Star Wars initiative as a way to develop systems for protecting U.S. missiles against a Soviet nuclear attack or counterattack.

If the United States ever builds an effective space-based missile defense system capable of limiting a Soviet attack, it will amplify the belligerency and recklessness of those policymakers who wish to fight and win a nuclear war. But even if research never leads to a working system, the Star Wars research initiative itself is a threat to peace. The Soviets view the research as a promise by the United States to "break out" of the 1972 Anti-Ballistic Missile Treaty. Hawks in the USSR are already using Reagan's commitment to Star Wars to justify their plans for new offensive and defensive systems, just as American hawks have used alleged Soviet research on particle beams to win support for the U.S. program.

In a cold war in which your opponent's *perception* of your military strength may matter more than your actual strength, research and development for future weapon systems serves, more than any other factor, to destabilize the military balance of power.

The ultimate significance of the Star Wars program is that the Reagan administration is using microelectronics and advanced computing technologies to embark on a no-holds-barred drive toward absolute nuclear superiority over the Soviets. In the words of the late Senator Richard Russell, "I have often said that I feel that the first country to deploy an effective ABM system and an effective ASW [anti-submarine warfare] system is going to control the world militarily."

■ Just as advances in military microelectronics technology have taken the combat zone into space, new developments in anti-submarine warfare are expanding the strategic arms race throughout the world's oceans. Today the United States operates a worldwide underwater surveillance system, in which high-speed computers analyze sound, radar, and magnetic data to find, track, and if need be target submarines at great distances. Soon submerged Soviet missile-carrying submarines may be vulnerable for the first time to an American preemptive attack.

Although the Soviets lag far behind the United States in the development of the compact, high-speed computers necessary for anti-submarine warfare, American policymakers expect the Soviets to modernize their anti-submarine forces too. To counter the perceived Soviet threat to America's existing submarine fleet, they are building submarines and missiles capable of hitting Moscow from virtually any ocean on the planet. To permit submarines to hide deep below the surface, they are developing communications technologies, such as the Extremely Low Frequency (ELF) system and the proposed satellite-borne blue-green laser, which can penetrate deep under the ocean waters with orders to launch their deadly cargo.

In the final year of the Carter administration, policymakers formally adopted a military doctrine that had been creeping slowly into American war planning. Instead of merely threatening the Soviets with massive destruction, the United States now plans to fight to "win" any nuclear encounter. This means that the United States is working toward the capability of carrying out a first strike effective enough to blunt Soviet retaliation.

To knock out Soviet land-based missiles in their silos and commanders in their bunkers, the Pentagon is working hard to develop nuclear weapons which can come within a few hundred feet of targets halfway around the world. Just as the Minuteman guidance system depended upon the first generation of integrated circuits, so the advanced guidance systems of the MX missile, the Trident II submarine-launched missile, and the Cruise and Pershing intermediate-range missiles all rely upon much more sophisticated microelectronic circuitry.

To be able to launch a nuclear first strike effective enough to limit Soviet retaliation, the United States is developing numerous missile guidance technologies, based upon three-dimensional gyroscopes, navigational fixes from stars and satellites, and "smart" weapons which home in on their targets. The Pershing II missile, now being fielded in Europe, contains a guidance computer that can instantaneously compare radar reflections of the earth with an electromagnetic map programmed into its memory. The Air Force is working on similar but more complex systems designed to enable intercontinental ballistic missiles to hit a kopek in Moscow.

The United States, merely by working on missile systems that might someday be accurate enough to carry out a preemptive attack on the Soviet Union, is actually bringing us closer to the nuclear brink. The Soviets will undoubtedly respond by modernizing their own weapons. More important, if they determine an American first strike to be a realistic possibility, they will place their missiles on a hair-trigger alert, ready to launch at the first—perhaps inaccurate—signs of an American attack.

The Pentagon strategists envision a limited, continuing nuclear conflict, in which the superpowers trade a series of carefully aimed nuclear salvos. But nuclear war, once launched, cannot be controlled. The new strategy does not lessen the chance of a global holocaust. In fact, the military's preparation for protracted nuclear war increases the likelihood of global war by creating the illusion that it can somehow be limited. The Pentagon is spending several billions of dollars each year fostering that dangerous illusion.

If a nuclear war begins, the President and his top aides will be rushed to a specially equipped Boeing 747 known as the National Emergency Airborne Command Post (NEACP, pronounced "Kneecap"). The command post is a combination "Führer bunker" and airborne electronic nerve center. It contains a presidential suite, bunks for the staff and crew, and food and fuel to keep the plane aloft and functioning for seventy-two hours.

Unlike the flying mansions of Arab princes, the NEACP is outfitted as an electronic palace, with special video game–like displays to track the progress of a nuclear war. The plane also contains coding systems, a five-mile reel of antenna wire for broadcasting messages at very low frequencies, and a variety of special military two-way radios. It can quickly establish contact with the NAVSTAR navigation satellite system, which contains equipment for pinpointing nuclear explosions anywhere on or above the earth's surface. If the President and the Secretary of Defense make it to the doomsday plane before Washington is vaporized, they will cruise comfortably outside the range of Soviet weapons while nuclear war rages on earth.

The Pentagon is constantly adding new electronic gear, such as terminals for new generations of nuclear-survivable communications satellites, to the NEACP and other doomsday command and relay aicraft operated by the Air Force and Navy. The Navy is replacing its TACAMO ("Take Charge and Move Out") planes, the sole function of which is to issue launch commands to missile-bearing submarines, so that they can cover the vast stretches of the Pacific now cruised by its new, powerful Trident submarines.

The Strategic Air Command, which controls America's nuclear bombers and land-based missiles, operates its own fleet of airborne command posts, code-named Looking Glass. Since 1961 at least one of these planes has been airborne at all times, 24 hours a day, 365 days a year, carrying an Air Force general whose sole task is to be ready to preside over Armageddon. In an eerie preparation for nuclear war, the pilots of these planes train to fly with a patch over one eye, so that even if one eye were to be blinded by the brilliant light of a nuclear explosion, they would still be able to function.

Light, however, is the least of the problems faced by those responsible for designing a command and communications system capable of surviving in a nuclear war. Each nuclear blast generates a massive burst of electromagnetic energy. In fact, a single nuclear warhead detonated fifty miles above North America would black out the entire continental United States, disrupt most radio and telephone communications, and permanently destroy all unprotected electronic circuits, including chips, in its path. Similarly, a blast in space would immediately disable unprotected satellites.

Pentagon and Energy Department researchers are therefore investing hundreds of millions of dollars each year on the research, development, and procurement of systems designed to resist electromagnetic pulse (EMP) and other nuclear weapons effects. Though such "hardened" devices are too expensive for most civilian functions, the Pentagon is gradually building EMP-resistant systems for most of its missions. One communications satellite has actually been exposed to the radiation of an H-bomb explosion in an underground nuclear test chamber; but most systems are checked out in test beds which simulate electromagnetic radiation. There is no way, however, to test critical electronic systems in a realistic nuclear war situation. America's entire strategy of planning to fight a nuclear war like a chess game is dependent upon technologies that may not work.

■ Microelectronics technology has not only made global nuclear war possible; it has thoroughly transformed the way conventional wars are fought. Modern-day electronic warriors frequently do not even consider themselves to be soldiers. They have become technicians, controlling a complex machine of which few can see all the parts.

In another era, a soldier looked at the man or woman he killed. Today, a soldier is frequently not only removed from the physical act of confronting the victims of violence; he may be removed from contact with traditional weapons entirely. Even the infantryman may forsake his rifle for a low-power laser to designate a target for a "smart" bomb, artillery shell, or missile that homes in on the laser's beam. Today's soldier may be stationed in an air-conditioned office or bunker hundreds or even thousands of miles away from the war zone. Still, he is able to control weapons which reach their targets faster and more accurately than ever before.

With modern telecommunications and computer technology, decision making, not just responsibility, is moving further from the scene of battle. In Vietnam, these new technologies permitted civilian planners in the Pentagon, led by Robert McNamara, the former president of the Ford Motor Company, to run the war like a multinational corporation.

Computerized reports, relayed via satellites, created the illusion in Washington that policymakers knew exactly what was going on. But even if McNamara had had today's more advanced high-tech equipment at his disposal, he would not have had a clearer picture. Commanders and intelligence officers transmitted the same inaccurate data that they fed McNamara on his occasional visits to Southeast Asia.

In Vietnam, the United States also first tested the automated battlefield. Faced with soldiers who were unwilling to fight and a public weary of the steady reports of American casualties, the military replaced ground combat troops with remote sensors linked to computerized equipment operated by technicians in faraway air-conditioned command posts. Like maniac deerhunters, American forces in Vietnam fired upon anything that moved. Electronic sensors helped them detect movement at a great distance, but they still could not distinguish between mice and men, water buffalos, or tigers. The Viet Cong disabled "people-sniffing" electronic devices simply by placing bags of urine nearby.

In southern Laos, American bombers dropped tons of

explosives on the Ho Chi Minh Trail system in an attempt to halt bicycle convoys. Small electronic sentry devices transmitted reports of activity to air bases in Thailand, but American pilots never really knew if their bombs hit a military target.

Experience and research have since increased the reliability of electronic battlefield devices, but a fundamental problem still limits the value of electronic intelligence and targeting methods in counterguerilla warfare: There is still no way for an electronic sensor to survey the workings of a person's heart or mind, to separate friend from foe.

The absolute failure of the Israeli armed forces and the Reagan administration to pacify Lebanon illustrates that weapons alone, no matter how sophisticated, cannot win wars. While individual pieces of high-tech equipment often prove useful in conventional warfare, an automated military cannot succeed where a traditional, human-based army fails.

■ In its modern, electronic form, war never ends. Superpower aircraft, ground stations, and ships in all corners of the globe are engaged in a daring "game" of electronic chicken—a constant battle to collect electronic intelligence that is not only part of American and Soviet preparations for nuclear war but is capable of triggering global conflict itself.

The only way to "fingerprint" the air defenses of a potential adversary is to test them by challenging them, head on. This is because the United States, the Soviet Union, and other modern military powers turn on their most important tracking systems only when they perceive a threat. To detect Soviet radar sites, as well as monitor frequencies and techniques, U.S. planes regularly approach Soviet air defense zones which, like their

American counterparts, extend well beyond the boundaries of the USSR.

In September 1983, the downing of a South Korean airliner, KAL Flight 007, by a Soviet air-to-air missile briefly raised the cloak of secrecy around the constant U.S. effort to map and analyze Soviet electronic defenses. What led the Soviet Union to risk international condemnation and shoot down a commercial airliner carrying 269 passengers and crew members during peacetime? How far up the American and Soviet command chains did news of the incident travel before the deed was done? Was Korean Air Lines Flight 007 on a deliberate spying mission?

Although it is likely that the mystery surrounding the events that plunged KAL 007 into the icy waters of the Sea of Japan will never be cleared up, U.S. intelligence agencies clearly knew at the time that the Korean 747 had penetrated Soviet airspace. In fact, U.S. electronic eavesdroppers used the confrontation to gather valuable data on both the general condition and specific methods of the Soviet air defense system in the region.

As the tapes of the incident revealed, U.S. intelligence stations in northern Japan routinely listen in on conversations between Soviet fliers and their superiors on the ground in Siberia. That capability, particularly if maintained in a period of international tension, gives American forces knowledge which can easily be exploited during actual battle. In fact, some former U.S. intelligence officials criticized the Reagan administration for disclosing too much information about our eavesdropping activities in its campaign to win international condemnation of the Soviet downing of KAL 007. Thus, the Soviet military may have obtained equally valuable information from the tragic episode, enabling them the better to conceal their communications in the future.

The secret and sometimes deadly electronic intelligence war has been going on since the end of World War II. With the rise of long-range missile systems in the 1960s, however, its importance as a tool for fighting nuclear war declined. Today, though, with both superpowers developing cruise missiles that operate at altitudes and speeds similar to those of manned aircraft, air defense systems are once again rising in importance.

Israel's aerial victory over Syria in the 1982 Lebanon war, though it did not involve nuclear weapons, illustrated the military significance of "electronic chicken" intelligence gathering. For months prior to the conflict, Israel sent unmanned reconnaissance aircraft over Syrian air defense positions. When the Syrians fired on the planes, Israeli officials quickly condemned them for attacking unarmed aircraft. But the Syrians were doing exactly what the Israelis wanted.

By the time the Israeli Army moved north, Israel had developed a comprehensive profile of the Syrian air defense system. When the Syrians turned on their radar to defend against armed Israeli aircraft supporting the invasion, Israeli missiles homed in on their signals and destroyed the entire surface-to-air missile system, giving Israeli planes a free ride over the battle area.

A later U.S. effort to collect its own intelligence on Syrian positions in Lebanon illustrated, however, how intelligence missions can escalate into armed conflict. When Syrian forces fired on a U.S. reconnaissance aircraft in December 1983, the United States launched a "retaliatory" air raid. This time the Syrians downed two U.S. planes, killing one American flier and capturing another.

In fact, the cold war is filled with incidents in which U.S. intelligence craft drew fire. Since 1950, a British newspaper reports, at least 27 U.S. aircraft have been forced or shot down and 60 others attacked while on electronic or photographic reconnaissance missions;

more than 139 U.S. servicemen have lost their lives as a result. In 1969, for example, North Korea shot down a U.S. Navy EC-121 over the Sea of Japan. Though they will never admit it, the Soviet defenders who brought down Korean Air Lines Flight 007 in 1983 probably mistook it for a military RC-135 reconnaissance plane, at least one of which flew through the area earlier the same day on an electronic intelligence mission.

None of those incidents has triggered global thermonuclear war, but at times when tensions are high, a similar attack could easily escalate into war. What if Soviet radar, set to trigger an automatic response to an attack warning, were to mistakenly identify a U.S. ELINT plane as a cruise missile? What if the United States decided to retaliate for the downing of another unarmed airliner? Just as Israel started its Lebanese war long before its troops invaded, World War III is already here, in the form of electronic feints and counterfeints.

■ It is well past midnight on the fifteenth floor of the Las Vegas Hilton Hotel. The Air Force colonel shares one last drink and remarks: "Westinghouse is just down the hall. You'll find a bunch of drunken folks in there. . . . It's a good show!" The colonel is right. It turns out to be a very good show. Down the hall a small army of military brass and civilian electronic warfare company executives and specialists are mixing business with pleasure. Pleasure is clearly winning out for the time being, but their business—how to win the next war—is very serious indeed. Gathered here are engineers and fighter pilots, military men and marketing men, university professors and spies, all of whom are members of an obscure but powerful fraternity known as the Association of Old Crows.

Visiting with the Crows leaves an outsider with a

disconcerting sensation. If these are in fact the warriors of the future, casting a glance around a room full of Crows makes one realize that there are very few Luke Skywalkers or physically fit astronauts here. It's obvious that the new all-electronic warrior won't be a grizzled, combat-hardened fighter or athletic young soldier. These are white-collar soldiers.

During World War II, the British Royal Air Force called the officers assigned to jam German radars the Ravens. Not to be outdone, the Americans nicknamed their electronic warfare officers Crows. After the war, veteran Crows formed an association, but until the Vietnam War, few in the Pentagon considered the work of the Crows important. Vietnam taught the military a lesson, however: Modern warfare is based upon electronics.

The Crows' common purpose in life is the advancement of "electronic warfare," defined by the military to include only a small segment of high-tech weaponry: radar-deceiving countermeasures, electronic intelligence, and related work. Officially sanctioned by the Department of Defense, the Association's annual convention offers Crows in industry and government the opportunity to cross-pollinate the seeds of new generations of high-tech weapons. "Hospitality suites" like the one sponsored by Westinghouse breathe fresh new meaning into Dwight Eisenhower's warning to beware the "unwarranted influence" of the military-industrial complex.

The magnitude of the industrial enterprise and technological base needed to wage modern wars dictates that entire societies be organized to support their preparation. As the political scientist Harold Laski has pointed out: "In the new warfare, the engineering factory is a unit of the new army, and the worker may be in uniform without being aware of it." Seventy percent of all federal-sponsored research and development funds are spent by the Pentagon and Department of Energy on "national defense," while more than one fifth of America's engineers work on military projects.

Indeed, the industrial arm of the military has grown so large in the United States that its needs now dominate the armed forces. And electronics is beginning to dominate military industry. Electronic weapons currently account for 30 percent of all Pentagon weapons purchases; analysts expect that figure to rise to 50 percent by 1990.

Military contractors and their legions of high-tech warriors work incessantly to build increasingly sophisticated high-tech weapons, frequently sacrificing reliability and effectiveness while inflating costs. For example, the Air Force has loaded its fighter aircraft with advanced guidance, communications, and countermeasures hardware—electronic equipment that is produced in relatively small quantities to unusual specifications. Unlike mass-produced electronic goods, such as microcomputers or telephones, the black boxes and displays purchased by the Pentagon are steadily increasing in cost.

James Fallows, whose *National Defense* is the bible of the Military Reformers, who consider "gold-plated" weapons the Achilles heel of American power, found that the price tag for electronic systems on a typical U.S. fighter plane rose from about $3,000 immediately after World War II to about $2.5 million a quarter-century later. And costs have jumped much further since. But as funds for buying planes are finite even during the Reagan presidency, the skyrocketing cost of complexity has forced the Air Force to reduce its arsenal of tactical aircraft.

The Military Reformers argue that the increased capability of jets equipped to launch "smart" missiles, confuse enemy radars, and navigate automatically does not make up for the limits that costs have placed on quantity. Despite computer simulations which projected that an advanced $30-million fighter jet, such as the F-14 or F-15, could down as many as seventy-four simpler F-5 fighters, the Air Force has determined in mock air battles staged near Las Vegas, Nevada, that the kill ratio is narrowed to 5 to 2 in confrontations involving multiple aircraft.

Further, the more complex a system is, the more likely a weak link will cause failure. Two Mariner spacecraft, en route to Mars, were lost in space due to simple programming errors. On one, a computer program contained a period where there should have been a comma; on the other, a key program was missing the word "not." In an era when both superpowers rely on complex computerized command and control systems to warn of an incoming nuclear attack, the consequences of computer failure—whether due to a hardware breakdown, a programming error, or unanticipated interference—may be catastrophic.

On June 3, 1980, a defective computer chip triggered a display at the Strategic Air Command's Nebraska headquarters, showing that Soviet submarines had launched missiles against the United States. SAC immediately scrambled its B-52 bomber crews and ordered them to start their engines, but conferring commanders caught the error before any left the ground. World War III was prevented by information from sensors that did not independently confirm the attack, and more important, by the actions of human beings in the decision-making "loop." In a period of greater international tension, however, a similar false warning could have triggered a more serious, irreversible response.

The June 3 incident sparked a congressional investigation, disclosing to the public for the first time that false alarms are uncomfortably common. During 1979 and the first half of 1980, there were 3,703 false indications of a possible Soviet attack on North America. Four of those reports, including the June 3 incident, qualified as serious false alarms. On October 3, 1979, for example, a radar station in Oregon detected a decaying rocket body crashing toward earth. A month later, the computers reported a mass Soviet attack when a computer tape containing data simulating such a raid was inadvertently fed into the system. Such errors were serious, requiring human analysis of conflicting data, because they were not anticipated. The warning system computers, therefore, could not have been programmed in advance to ignore the indications of an attack.

Today, the United States not only relies on the "man-in-the-loop" to confirm an incoming attack, but the President must approve any American nuclear launch. However, in a surprise attack on the United States, there might be just a few minutes between the detection of Soviet missiles and the flattening of Washington. There might not even be enough time to wake the President. For this reason, many strategic planners advocate the adoption of a "launch-on-warning" strategy, in which U.S. missiles would be fired immediately and automatically when the North American Aerospace Defense Command's (NORAD) computer determined that the United States was indeed under attack.

U.S. leaders refuse either to endorse or to rule out launch-on-warning, but it doesn't matter. The armed forces strategic warning network is so complex that both the NORAD commander and the President are already dependent upon NORAD's computer.

To process data from that network, the Air Force has hired high-tech contractors to write "expert system" software, programming machines to make "decisions" based upon preestablished logical rules and data. *De-*

fense Electronics magazine, formerly the mouthpiece of the Old Crows, says that such programs "would be a substantial aid to space defense officers and field commanders, not to mention the sweaty-palmed officers manning the North American Aerospace Defense Command's early warning scopes."

Placing "intelligence" in systems that control hair-trigger weapons of destruction is certainly the height of suicidal folly. Critics, including members of Computer Professionals for Social Responsibility, liken the Pentagon's approach to borrowing to get out of debt. To solve problems which have been created by the warning system's complexity, the Air Force is developing even more complex computer systems. Just as the debtor merely ends up in greater penury, the future of the world is becoming dependent upon even more intricate computer systems. By definition, these systems are untried and untested. Millions of simulations can be run, yet there is no way of ever knowing if fatal bugs remain, or how such phenomenally complex systems will respond during a crisis, particularly in the face of electromagnetic pulse and other nuclear effects.

In the long run, there is no technological solution to the problem of missile warning. The chance of nuclear war by mistake will continue to increase unless the arms race is halted.

■ The military industrial complex, however, is structured to propose only technological solutions. Since automated fighting systems—whether for Vietnam-type combat, Star Wars, or missile warning—are unreliable, the Pentagon wants to invest the superior intelligence of *Homo sapiens* into its machines.

In 1983, the Pentagon's Advanced Research Projects Agency (ARPA) launched the Strategic Computing initiative, a $600 million research program designed to develop machines that emulate the reasoning functions of the human mind. The program combines several of the most advanced research areas of computer, chip, and software technology.

Unlike previous ARPA research programs in computer science, Strategic Computing is directly linked to military goals. The initiative has different "intelligent" weapons in mind for each service of the armed forces. For the Army, there is an "autonomous land vehicle" (killer robot); for the Air Force, a "pilot's associate" (R2D2); and for the Navy an "intelligent battle management system" (an admiral's adviser).

The autonomous land vehicle is planned to combine expert systems with vision systems—two areas in which both the military and commercial researchers have already invested dozens of years and millions of dollars. A fully operational weapon is planned for 1993. The robot is supposed to make it own maps by 1989, and computer scientists have been instructed to endow it with the ability to identify trees, bushes, grass, and rocks, and to infer the presence of hidden objects by 1992.

Pentagon scientists hope that the pilot's associate will be able to recognize more than two hundred words of speech, as well as speak more than one thousand human-quality words. It will also be able to recognize terrain, tell the difference between friends and enemies, offer tactical advice, and control weapons. The admiral's adviser is supposed to display a detailed picture of a battle area, estimate the likelihood of various enemy actions, and offer courses of action to the fleet commander. The battle management system will be programmed to learn from its successes and failures—if it survives.

In the long term, the Pentagon plans to apply the results of the Strategic Computing program to the control of space-based weapons, including the proposed Star Wars system. These satellites will need high-speed, autonomous computers to act quickly enough to knock out Soviet missiles in their "boost" phase. However, a space platform capable of launching devastating missiles or beam weapons without instruction from the ground would be the incarnate version of the fictional "Doomsday Machine" that ended the world in the movie *Dr. Strangelove.*

Perhaps more than any other factor in the electronic arms race, the Strategic Computing initiative threatens to propel events truly out of human control. The specter of autonomous fighting machines is a clear case where the United States is dictating the pace of superpower arms competition. The Soviets, who lag far behind in computer technology, are certain to feel compelled to respond. Thcy will no doubt shortly embark on a course that will lead to the deployment of their own generation of "intelligent" weapons.

Is it possible to pull back from the abyss of nuclear war orchestrated by machines? The late Senator Henry Jackson, considered a hard-line hawk, made an ambitious proposal before he died. Along with Armed Services Committee colleagues Sam Nunn and John Warner, Jackson suggested the formation of a permanent Soviet-American Joint Consultation Center. The purpose of the Center would be to prevent war by accident or misunderstanding. Jackson wrote: "I envisage it as consisting of a jointly operated central building providing working space and conference rooms for both [Soviet and American] staffs, and adjacent nationally controlled buildings, one run by the Soviets and one by us. These would give the staff of each side a chance to confer privately and enable them to be linked to their respective capitals by ultra-secure, unilaterally controlled communications." Each side's staff would utilize its own intelligence and data analysis resources.

The Center would not test new technologies, but it would use recently developed information technologies in a new way. Both sides would essentially be forced to explain suspicious events, detected, recorded, and relayed via electronic systems, before they escalated into global confrontations. Jackson believed that such a Center would defuse crises; he also felt that it would create an environment in which representatives of both superpowers would become accustomed to working together on a daily basis.

Since the early 1960s, when the United States and the USSR signed a treaty banning nuclear weapons tests in the atmosphere, high-tech intelligence systems have been an integral feature of arms control. Because the superpowers' armed forces are unwilling to accept most forms of on-site inspection, arms controllers have relied heavily upon reconnaissance satellites and other forms of electronic intelligence to verify treaty compliance.

However, most intelligence-gathering equipment has a second purpose: to pinpoint potential enemy targets. Reconnaissance satellites, for example, are also a key part of offensive systems. Those who propose arms control are of course reluctant to place any limits on their high-tech tools. Still, it is possible to place enforceable technical restrictions, such as a requirement that communications links have built-in twenty-four-hour delays, which would preserve the peacekeeping function of a satellite while limiting its offensive function. For example, equipment for detecting nuclear explosions, on-board NAVSTAR satellites, could be designed to supply arms controllers with vital information on nuclear testing without promising the practitioners of protracted nuclear war the data they need to retarget weapons in the midst of a limited holocaust.

Treaties can also control new weapons systems long

before they are built. Even a nuclear freeze would soon become meaningless unless research and development, a sacrosanct process in which the technical establishments of each superpower continually dream up more ghastly weapons, is also restricted. By the time a new missile, bomber, or submarine makes it to the bargaining table, it has a momentum of its own. Its sponsors have spent billions of dollars on development, and the armed forces of the other superpower are likely to have spent billions on a counterweapon.

One of the most important arms control agreements ever negotiated between the United States and the USSR, the 1972 Anti-Ballistic Missile Treaty, specifically permitted each side to continue research and development for new systems. Since the treaty was signed, the U.S. military has spent billions of dollars on missile defense research and development. Now, a little more than a decade later, the treaty is virtually worthless in the face of the Reagan administration's "Star Wars" research initiative.

Still, if national leaders are willing to negotiate controls on applied research and development, work on major new weapon systems can be identified and restricted well before they are ready for deployment, without hampering normal scientific progress. The task of monitoring compliance with agreements regulating research and development can be eased by limiting the use of academic and industrial secrecy.

■ The trajectory of U.S. policy offers little reason for optimism. In fact, it is absolutely critical to look beyond official pronouncements that the electronic arms race is necessary to deter war. History shows that arms races lead inevitably to war or to economic disaster. Why should this arms race prove any exception?

And the stakes this time, even in conventional war, are thousands of times higher than ever before. Even the 1973 Arab-Israeli War, one of the first "unlimited" tests of modern weaponry, consumed $10 billion worth of weapons and supplies in the space of two weeks. The potential levels of destruction continue to escalate exponentially.

Microelectronics-based technologies have become the driving force behind the arms race. A constant electronic cold war rages between the superpowers as they probe and test each other's defenses. Each measure prompts a new countermeasure. Electronic warfare systems, which become obsolete within the first year or two after they are deployed, are replaced by new black boxes. The fielding of new electronic weapons often renders missiles, ships, and planes ineffective (against another superpower) long before they are deployed, further accelerating the pace of weapons development.

Hair-trigger warning systems, "intelligent" weapons, electronic chicken, Star Wars, nerve centers, survivable satellites, remote command and control—the elements of the electronic arsenal are all converging in a science fiction nightmare. Unless we as a society place the genie of high-tech weaponry back in the bottle, the nuclear button may soon be firmly and perhaps irretrievably planted in the "hands" of an unthinking automaton thousands of miles overhead.

Chapter Three

OUR FRAGILE FREEDOM

■ ■ ■ ■ ■

Outside, the sun is beating down. Inside, there is a perceptible air-conditioned chill. The room is windowless. Lighted fluorescently, it houses only a minicomputer, a printer, a bank of tape recorders, and an operator's console. A man wearing headphones sits at the console, listening intently and occasionally entering instructions on a keyboard.

The time is the final years of the rule of the shah of Iran; the place, Doshan Tappeh Air Base, near Teheran.

The shah's TM-4020 Telephone Monitoring System is simultaneously tracking 3,500 telephone lines. It notes which phones are placing calls and which are receiving them, creating a secret digital record of who is talking to whom, for how long, and in what sequence. The system also makes a record of every misdialed and dropped phone, busy signal, wrong number—mistakes of any kind that might reveal information about the dialer. When targeted lines call or receive, a tape recorder automatically switches on or the computer immediately alerts the operator to the connection. And all of this is undetectable.

Today, the shah is gone. No one is saying what happened to the critics of the regime who were caught in his electronic dragnet. But the technology lives on. In the United States, where a small Silicon Valley firm developed the TM-4020, progress in telecommunications and surveillance technology has already turned the shah's system into an antique.

■ At about the same period, the West German federal police produced an investigatory technique called Raster-Search, in which a computer sorts through millions of personal data files to list suspects matching the profile of a possible terrorist or troublemaker. The computer checks the 55 million entries in a social data bank that includes reports on individual health and income; it searches through the electronic files of the insurance industry, which cover 47 million people; and it peruses the data files of utilities, real estate agencies, and schools.

The first search may turn up tens of thousands of pre-suspects, but subsequent comparisons usually narrow down the list to a few hundred individuals most closely matching the profile. The computer not only provides the names of the suspects; it also generates a complete data file on each one. An anonymous German scientist, writing under the pseudonym of Marion Butner, says that any of the Raster-Search suspects can be detained, questioned, or even arrested by police solely on the basis of their computerized profile.

Though the ostensible purpose of the Raster-Search is to apprehend terrorists, Butner charges that "many political activists were caught up in the spider web of Raster-Search. In most cases the result was not prison, but rather loss of job, privacy and/or housing."

■ In the United States of the 1980s, high-tech surveillance systems that dwarf those of the shah and the German federal police are already in place. Laws to protect personal privacy have not kept up with advances in list-keeping, electronic eavesdropping, and information processing. Although the Bill of Rights is supposed to protect Americans against unreasonable search and seizure, microelectronics technology has rendered the Fourth Amendment obsolete.

The guarantee of individual privacy is more than an abstract ideal; it is the first line of defense against a police state. Many times in U.S. history, those in power have used personal information to suppress dissent. During the McCarthy era, for instance, both the government and private organizations punished people for their views and associations. In the early 1970s, however, advocates of civil liberties won new protections against the abuse of power when the public reacted angrily against the Watergate break-in.

Today, Americans live in a state of fragile freedom. Wide-scale repression is still unacceptable, yet surveillance technology has made tremendous strides during the past decade. High-tech systems have been developed that can and are being used to monitor our lives in ways never imagined by George Orwell.

■ The consumer credit bureau at TRW (formerly Thompson Ramo Woolridge), a firm better known for its close to $1 billion in contracts each year with the Pentagon and intelligence agencies, watches over the spending habits of 90 million consumers. The TRW system is constantly growing, and its operators are constantly updating the computer hardware and software that keep it ticking.

The most sophisticated elements of the TRW network are those that allow it to communicate with its clients. Since 1980, the company has operated a network called Datalink, which links TRW's data center in Anaheim, California, to its regional offices. Datalink is capable of handling about 35,000 reports per hour, or about 325,000 per day.

Among TRW's clients are some one hundred federal agencies. Just like merchants, public officials have direct twenty-four-hour-a-day access to the credit company computers.

Computerized lists like TRW's are a virtually inescapable part of modern life. Anybody who has ever received junk mail knows that his or her name—often misspelled—and address rests comfortably in the electronic memory of multiple data banks of unknown place and purpose. The average American is listed on computers belonging to most of the following: airlines, banks, car rental agencies, the Census Bureau, consumer research firms, credit bureaus, credit card companies, employers, health care agencies, insurance companies, the Internal Revenue Service and its local counterparts, law enforcement agencies, magazines, motor vehicle and drivers' licensing bureaus, politicians and pollsters, schools, selective service, Social Security, utilities, and voter registrars.

From those files, a public or private investigator can quickly draw an exhaustive personal profile on anyone at any time. Historically, when attempting to create a file on a suspected criminal or dissenter, investigative agencies have had to rely upon newspaper clippings, informers, and court records. Although a small amount of information could be obtained retroactively by searching various files, it has generally been necessary to identify the target of an investigation before initiating the monitoring process.

Today, however, it is possible to study the behavior of a person who has a clean legal slate, merely by linking data already collected and stored in computerized form. Without ever interviewing a target's friends, eavesdropping on his/her phone, or sending spies to meetings, investigators can determine the individual's political party, magazine subscriptions, air travel plans, financial status, and medical condition, as well as the identities of friends and acquaintances.

The more we rely upon electronic means of doing business, the easier it is for someone to record our business in their computer. Already, credit card companies such as Citicorp Credit Services combine their purchase records with demographic data in order to prepare detailed customer profiles, for sale to retail business. Markets and bookstores with computerized cash registers develop detailed records of what we read and eat. All that is needed to turn those records into an investigatory data base is a personal identification code, such as a credit card or check cashing number.

Some data banks are merely high-tech blacklists. For example, RentCheck Services, of Denver, Colorado, maintains a list of more than 1 million "problem" tenants. Landlords across the country turn in the names of tenants who they say have damaged property, skipped rent payments, or challenged a cleaning deposit deduction—the last category being entirely legal. Other landlords check on prospective tenants through RentCheck, denying housing to apartment-seekers blacklisted by the service. In most cases tenants do not even know why they are turned down, so they have no way to correct misleading information in the RentCheck files or to justify their past actions.

Despite the potential for abuse, data banks containing information on our personal lives are here to stay. Without them, public and private social programs, such as Social Security and health insurance, would be cumbersome, if not impossible. Bank credit cards, airline reservations, and magazine subscriptions would be unreliable, if not unavailable. Our long-distance phone bills may contain valuable personal secrets, but when we call to question a charge on our bills, most of us like the instantaneous response provided by the phone companies' computers.

The challenge, therefore, is not to eliminate data banks but to develop policies, legislation, and even technology that will guarantee or at least help to protect personal privacy.

■ Scott Robinson is a truck driver who was serving time briefly in Santa Clara County's jail for stealing video games. He said he wanted to be home for Christmas.

Since he was working as a "trusty," doing maintenance work in the booking section of the main jail, Robinson had access to the county jail's computer. When sheriff's deputies were looking the other way, he allegedly logged onto the machine and altered his release date from December 31, 1984, to December 5. A deputy, suspicious about Robinson's release date, discovered his scheme when he checked the jail's written records. Investigators still do not know exactly how far Robinson got on his electronic ride through the criminal records system—but he made the front page of several newspapers.

Robinson's exploits, as well as the highly publicized "break-ins" by teen-age computer hobbyists, illustrate how easy it is to get into large computer systems. The data banks of law enforcement agencies, credit bureaus, employers, and so on are essentially open books on our lives.

The threat of intrusion by hobbyists is exaggerated, since most break into computer files merely for the challenge of outwitting the security precautions. Usually they are uninterested in personal data, and pose no major threat to individual privacy. Occasionally, however, a professional computer break-in is detected, in which the persons or organizations gaining access are seeking financial gain. In 1982, for instance, police in Orange, California, not too far from TRW's computer center, confiscated thousands of credit files from a car repossession agency and a private investigation firm. The data thieves did not qualify as TRW customers, since the credit bureau makes its information available only to bona fide issuers of credit and to law enforcement agencies. But with teleprinters and ID codes, the intrepid investigators were able to pose as TRW customers for ten years before being spotted by the company.

In most cases, however, the data thief is an insider. Bank tellers, doctors, welfare workers, teachers—people in all kinds of work using personal data banks—can easily request data on customers, patients, clients, and students from their institution's computer. Sometimes the motive may be curiosity—"Why was my professor visiting the VD clinic?", or blackmail—"I'll tell his wife." But the information can also be used for financial gain or political influence.

Law enforcement officers with political connections can easily divert legally collected personal intelligence. In January 1983, the Los Angeles Police Department discovered that one of its detectives, Jay Paul, was running a $100,000 computer system from his wife's law office for Western Goals, the investigative arm of the ultra-right John Birch Society. Reportedly, Paul's computer records included data illegally copied from official police files. Most of the files remain private, but the Los Angeles *Times* said that Paul had a two-page dossier on an employee of the American Civil Liberties Union, Linda Valentino, listing employment, medical, and voter registration information.

Still, the greatest threat to personal privacy is posed by the authorized use of high-tech data banks designed to collect detailed political, personal, and legal information. Based on active surveillance, police and intelligence files can contain virtually any type of information. While no one objects to file-keeping on convicted criminals or those suspected of committing a specific crime, government agencies have often cast a much wider net of surveillance.

During the early 1970s, for instance, the FBI kept a list known as the Administrative Index, which secretly held information on at least 23,000 people who could have been detained by law enforcement agencies in a national emergency, even if they had committed no crime. Later, the FBI released its criteria for including a subject in the Index:

An individual who, although not a member of or participant in activities of revolutionary organizations or considered an activist in affiliated fronts, has exhibited a revolutionary ideology and is likely to seize upon the opportunity presented by national emergency to commit acts of espionage or sabotage, including acts of terrorism, assassination or any interference with or threat to the survival and effective operation of the national, state, and local governments and of the defense efforts.

The FBI wants to expand its National Crime Information Center (NCIC) data bank to cover people who have never been convicted or charged with a crime. It has proposed adding people with reputed organized crime connections, suspected terrorists, and associates of drug traffickers. The Secret Service wants to add to the files individuals it considers a threat to the President.

These changes conceivably might add only a few names to the NCIC's files; they could also open the door to a new explosion in electronic file-keeping. American police agencies, including the FBI, have historically had a difficult time distinguishing between terrorism and civil disobedience or even lawful protest, or between criminality and innocent association. Groups like the American Civil Liberties Union have objected to the NCIC expansion, and Congress has the power to block it. But it is likely that the FBI will get its way.

■ In the spring of 1979, Billy Carter (brother of President Jimmy Carter) and his business associates initiated discussions with the government of Libya, generally considered hostile to the United States, to handle the shipment of crude oil from Libya to America through the Charter Oil Company. The dealings of Carter, who had direct access to members of the National Security Council as well as to his brother, prompted a Senate investigation. To piece together the negotiations, a subcommittee of the Judiciary Committee subpoenaed toll and long-distance records covering one or two years of calls from Billy Carter's home phone, his office phone, and those of his principal associates. Even a year after the fact, the phone records provided investigators with a sketch of Carter's travels, confirmation of his contacts with the Libyan Embassy in Washington, and an outline of his relationships with his partners.

The modern telephone is no mere household gadget. The American telephone network is a series of complex interlocking computers, connected by microwave radio transmissions, satellites, and fiber-optic cables which intelligence agencies, law enforcement, the military, and private corporations can all use to obtain and manipulate information once considered safely private. Carrying voice, pictures, and computer data, the phone system routinely conveys intimate information about virtually every American citizen.

It is technically simple today to monitor long-distance phone conversations without going near either the communicating parties or the phone company itself. Most long-distance calls placed in this country travel by microwave—beams of electromagnetic energy that can be cheaply and easily intercepted without detection. To eavesdrop, one need only set up an antenna in view of the microwave communications link and then program a microcomputer (equipped with a decoding device) to scan the airwaves. The equipment is cheap, and legal to own. The techniques are in the public domain.

The simplicity of this method has led numerous publications and political figures to charge that Soviet agents in this country are conducting widespread telephone surveillance. It is technically possible for Soviet consular representatives in San Francisco, for instance, to intercept both vocal and computer communications from nearby Silicon Valley, though no one has proven that such snooping actually occurs.

Before AT&T recently revised its phone-signaling procedures, it was actually possible to listen in, via microwave, on long-distance calls to targeted telephones. Dialing information was sent in the same telephone channel that carried voice communication. When a computerized scanning device determined that a particular phone was calling or receiving, it could automatically trigger a tape recorder.

Today, nearly all dialing information travels in a separate channel from voice. Few organizations can afford the computers and personnel to sort through the large volume of telephone traffic, but the U.S. National Security Agency (NSA) specializes in this task. It is fitting that in the age of microelectronics, the U.S. espionage agency that dwarfs all others is the one that specializes in electronic surveillance.

The NSA, like the CIA, is not authorized to collect domestic intelligence. Yet it legally monitors a vast number of calls between the United States and foreign countries, providing summaries or transcripts of a small percentage of the calls to the FBI. The NSA's cloak of secrecy prevents us from knowing exactly how much tapping it does in the United States, but its activities elsewhere demonstrate its capability. We know that its Menwith Hill installation in England, the largest phone-tapping center in the world, can simultaneously scan 14,400 lines.

Still, in the telephone system, as with data banks, the greatest threat to privacy comes from those with inside access. Phone companies are empowered to observe calls, without even sounding a beep-tone, for the ostensible purpose of "improving" their own performance. With a court order, the FBI and other law enforcement agencies can tap phones to investigate crimes.

Using modern solid state switching equipment, a telephone-tapper need not bug a phone or "wire up" an exchange in order to intercept calls. Telephone company employees or police agents can tap a line merely by typing a few keys on a computer terminal at a remote location.

Strange as it may seem, some of the most extensive electronic eavesdropping conducted by U.S. intelligence and police agencies does not even involve any monitoring of "aural" signals—that is, those that can be interpreted by the human ear. The laws restricting eavesdropping are riddled with loopholes allowing for the interception of "nonaural" transmissions, including telephone dialing information, data communications, and facsimile transmissions.

David Watters, a former CIA official and telecommunications expert, describes the surveillance capability of a device used in state-of-the-art telephone exchanges: "The pen register can record a phone accidentally knocked off the hook. That tells you someone was home at a particular time. Many misdialed calls may indicate something about the caller's state of mind, distractions, memory problems, or if the misdialing happens at certain times of the day, there may be a drinking problem . . ." The pen register is based on technology that is now decades old. The modern computer, however, is in reality a giant and highly efficient pen register.

Today, the Department of Justice has a network of regional information centers, each with the responsibility for the computer analysis of toll-call information on behalf of state and local law enforcement agencies. Any American can trigger the special attention of the FBI or other law enforcement agencies merely by calling the wrong people or places.

Of course, law enforcement agencies are not the only ones to exploit dialing information. Employers with computerized private branch exchanges (PBX) can easily record the telephone activities of their employees. And with minor programming adjustments, they can flag calls to particular outsiders—such as competitors, reporters, or government agencies. Those calls can be automatically monitored, although the law still requires that the person at the other end be notified of any tapping by a periodic beep-tone.

■ In the era of the microprocessor, the form of electronic surveillance with the greatest potential for gathering data on the American public is the interception of data communications. Vast amounts of computerized data, including personal records, are constantly being transmitted over phone lines and cables, microwave relays, and satellite channels.

The task of actually intercepting and using this information, however, is an enormous one. To make full use of the data in the computers at TRW's credit bureau, for instance, the monitoring agency would need a computer center and network even bigger than TRW's, dedicated to just that task.

So a practical approach to intercepting computer communications would focus on particular targets, in search of specific information. The public record does not show how much of this type of surveillance the federal government in fact does. The National Security Agency undoubtedly has the capability to spy on Americans in this way, but it still might be easier for the FBI to phone a bank manager for a copy of a customer's transactions than to go to the NSA in the hope that it can cull that data from the phone network.

Incredibly, it is even possible for computer spies to copy information directly from a computer's low-level electromagnetic emissions. Unless a computer is tightly shielded, someone with a receiver and a computer for sorting through the signal can reconstruct information from a targeted computer. Since video screens emit electromagnetic signals, it is possible for a nearby observer to reconstruct the information displayed on a video screen, reportedly from as far away as 500 feet, without being detected. For example, one electronic security firm, as part of its sales pitch, drives a truck up to a computer installation, throws a wire over the power lines running into the building, and then shows the astonished employees what is printing out inside their offices.

The monitoring of computer emissions is difficult and expensive, but to foreign and industrial spies it can provide invaluable information. When they monitor the communications of agencies that encrypt their messages and files, they want access to a small stream of uncoded data. That segment can act as an electronic Rosetta stone, providing a translator's key to unlock a wealth of information.

As computers and surveillance equipment become cheaper and more widespread, the potential for this type of surveillance grows. Small computers, such as personal computers, are easier to monitor than mainframes, because the latter emit multiple signals which tend to interfere with each other. In addition, the higher-frequency emissions characteristic of today's ultra-fast computers make shielding less effective.

The National Security Agency operates a testing program through which it determines whether important government computers are adequately shielded against electromagnetic eavesdropping technology. The program is so sensitive that although the NSA rules whether machines are adequately shielded, it refuses to disclose its standards for making that determination.

Like computers, each television set also emits a low-level signal that can be monitored with a nearby antenna. From that signal, it is possible to determine the station to which the set is tuned. Broadcast pay-television stations send vans through residential neighborhoods with devices specifically designed to catch "pirates" who are receiving the pay-TV signals without permission.

So far, TV-rating services have considered this technique impractical for compiling viewer statistics. It is easier to sample cooperating homes across the country than to drive down everyone's street with a high-tech van. Still, any law enforcement agency wishing to profile the viewing habits of an individual or neighborhood could do so very cheaply and with little difficulty, merely by monitoring the secondary TV-set emissions.

Monitoring the viewing, dialing, and conversational habits, as well as the personal records of millions of people, is not only technically possible but is undeniably done in one form or another by both public and private agencies. But without the ability to process that information, their data collection would have little practical value. Without data processing, intelligence agencies would be like parade-watchers in New York City buried knee-deep in tickertape.

But data processing is exactly what computers do best. And the power and sensitivity of surveillance computers is shocking.

In order to sift through the huge number of aural conversations intercepted from microwave transmissions, the National Security Agency converts the signals into digital data. Nonaural dialing and billing information can be used to do more than identify potential investigative targets. For example, a California software firm has developed programs designed to help police agencies analyze the telephone habits of organized crime figures. In workshops staged periodically across the country, the company's instructors teach police agency students the techniques of "link analysis"—a method of determining informal patterns of organization—utilizing telephone toll-call information.

James Danowski, a university-based communications researcher, has developed computer programs that can analyze the telephone traffic patterns inside an organization. His software yields an accurate picture of who the leaders are and how they function—just from dialing information. These techniques can be easily applied to groups of phones in a ghetto or student community, he maintains, making invisible social networks visible and identifying key members of the community.

Former CIA engineer David Watters says that the first and second tiers of electronic surveillance involve digital tone scanning. He reports that computers are now on line that can do a statistical analysis of calling patterns, match that with other information, and develop a profile of caller behavior. Watters calls the notion that everyone's line is tapped an exaggeration, but he claims that intelligence agencies are analyzing a major portion of all long-distance calls by nonaural methods.

■ In 1977, a Palo Alto youth named Eric Hentzel submitted a phony entry in one of the Farrell's Ice Cream Parlour's birthday contests under the name of a fictitious character, Johnny Klomberg, age eleven. Six years later, the Selective Service System bought a magnetic computer storage tape listing 167,000 names of boys born in 1965 and 1966 who had participated in Farrell's contests. The System's computer compared the Farrell's list, as well as school records and other files containing the names of teen-age boys, with its draft registration records. In 1984, the Selective Service System sent Johnny Klomberg a postcard reminding him to register for the draft.

The computerized comparison of records in two or

more data banks, known as "matching," is likely to become the leading means of analyzing computerized surveillance data. Initiated by the federal government in 1977 and even earlier in some states, there are presently over 250 state and federal projects conducting such comparisons.

Matching, potentially a severe threat to the privacy of every citizen who pays taxes, has not generated widespread concern so far because it has been used to detect fraud, to track down fathers avoiding child support payments, and to identify public employees in default on student loans. The Justice Department is checking for dead people and double-registrants on the voter lists of Chicago and its suburbs.

The Internal Revenue Service is taking the matching approach a step further. Using public sources such as phone directories, vehicle registration records, and Census aggregate income data, as well as information from private marketing firms, it is attempting to estimate household income. If a household's reported income does not match the computer's projected figure, the IRS examines or audits the taxpayer's individual records.

Taken to their logical conclusion, individual matching projects could be combined to create one massive government data bank, comprising Census information, credit data, tax records, Selective Service files, voter registration, automotive registration, school records, and almost any other type of personal interaction.

Matching is indeed attractive when it catches a physician who has double-billed Medicaid, but in the long run it is dangerous and possibly even counterproductive. Like the Farrell's Ice Cream list of potential draft registrants, data files matched by government agencies frequently contain inaccurate or out-of-date information. Sometimes, two data banks use different definitions. For instance, a college student listed by a voter registrar or car insurance company as part of his parents' household may file his own tax return. Furthermore, as the scope of matching becomes known, those who seek to defraud the government as well as libertarians who object to the matching in principle will refuse to participate in vital government data programs. Already large numbers of people are refusing to cooperate with the Census.

■ High-tech methods for collecting, storing, and processing personal information would shock the framers of the Bill of Rights. "There is no significant difference," David Watters told Congress several years ago, "between electronic broadband interception practices and the early practice wherein agents of the Crown of England, during colonial times, armed with general warrant documents or writs of assistance, could plunder at random through the homes, offices, and effects of citizens, and could read, examine, or carry off any document or property thought to be in the interest of the King."

In the Bill of Rights, the Fourth Amendment rarely gets the publicity of the First, Second, and Fifth, but the text is well worth recalling here:

> The right of the people to be secure in their persons, houses, papers, and effects, against unreasonable searches and seizures, shall not be violated, and no Warrants shall issue, but upon probable cause, supported by Oath of affirmation, and particularly describing the place to be searched, and the persons or things to be seized.

Unfortunately, too many Americans consider the Fourth Amendment a mere legal technicality, since seemingly guilty criminals have used it to escape punishment. Some people believe that those who insist on personal privacy must have something to hide; and some feel that surveillance is always aimed at the proverbial other guy.

Privacy is a basic human right, obtained with the blood of the founders of this country and struggled over ever since. But privacy is more than a value in itself. Without protection against unreasonable searches and seizures, physical or electronic, a society lays itself open to tyranny.

The tools of computer and electronic surveillance, in the hands of centralized institutions such as the federal government or large financial institutions, provide those in power with an enormous amount of information about our actions and even our thoughts. Armed with that information, they can control and manipulate us. As tyrants gather more data, they need rely less on random force to achieve their ends, but their power is more absolute. The most effective authoritarian state is one in which the bonds of control are barely visible.

The government has the power to fire (or cause to be fired), arrest, or even execute dissenters. Without accurate information on leadership patterns, however, its repression becomes random, arbitrary, and less effective.

Surveillance provides would-be dictators with the information, such as data on the political structure and personal relations within protest groups, to undermine dissent. Though the mere knowledge that police agencies have such data intimidates many potential activists, personal information can and has been used to foster conflict within protest movements. The FBI's COINTELPRO operation helped destroy the growing political effectiveness of the Black Panthers in the 1960s, for example. FBI-sponsored provocateurs spread stories, some false but all plausible, within the organization, provoking bloody, destructive internal fights.

The history of the United States has been one of intermittent struggle between those who would control the population and those who share the spirit of the authors of the Bill of Rights. Sometimes, as in the Watergate scandal, our rights were reaffirmed. Other times, as during the Red scare of the 1950s, our freedoms were overrun. Today, American law enforcement agencies could easily combine West Germany's Raster-Search with a telephone monitoring system more sophisticated than the shah's. Only the current political climate prevents the forces of repression from applying their latest surveillance technologies against dissent.

But control can also take a much more subtle form. Already mass marketers and politicians use personal information, not to punish us, but to manipulate us. They cull data from market research firms and pollsters, as well as their own direct contacts with customers and constituents, to lump us into classifications of people expected to respond to particular messages.

Information Resources, Inc. (IRI), is a market research firm that monitors the TV-viewing and buying habits of fifteen thousand American households. Each cooperating family has a microcomputer attached to the back of its television set, noting the shows and commercials it watches. When a member of the household goes out to buy groceries, automatic scanners at cooperating supermarkets record exactly what he or she buys. IRI's computers then match the store and viewing data for each family.

IRI tells its clients, advertisers of consumer goods, whether a household that views a particular TV commercial is statistically more likely to buy the product. They in turn use IRI's reports to figure out how to manipulate the American public, not by producing better products, but by varying the timing and content of TV ads. Armed with such information, advertising agencies learn which TV star can best sell a computer to a particular audience rather than how to make machines more reliable. They figure out how to sell brand-name aspirin for much more than the generic product, and develop strategies to convince parents that costly, ecologically unsound disposable diapers are more "loving" than cloth.

Politicians use polling data exactly the same way. They find out what each constituency wants to hear, then they tailor their messages to each audience. Some may actually lie or change their views for political expediency, but in most cases candidates and officeholders merely select what they say to avoid negative reactions.

Public officials use their own legally collected computer files to manipulate us as well. Once an officeholder knows our interests and views, he or she can tailor apparently personal form letters to us. In Silicon Valley, one congressman sends out regular, personally addressed letters to all those constituents who urged him to support the House nuclear freeze resolution. He sends different letters to freeze opponents, and still other form letters to constituents on his lists for Central America policy, Social Security, environmental affairs, and so forth. Without lying, but by classifying and selectively informing voters in our area, he is able to build up a much more positive image than his full record would project.

■ Individuals can protect their privacy by avoiding behavior that is automatically tracked by commercial computer systems. They can use cash, avoid stores with scanners, and call from pay phones. Appliance buyers need not fill out the marketing questionnaires that accompany warranties. And no one has to enter the Farrell's birthday contest. But such privacy may still be an illusion. People who always use cash instead of checks or credit cards are so unusual that law enforcement agencies flag them as suspicious characters whose behavior can only be explained as a conscious attempt to avoid surveillance.

The operators of computerized data banks can limit the unauthorized review or modification of personal information in their files through techniques such as call-back and encryption. Call-back simply means that a remote computer user cannot directly call into a computer system. Rather, the host system receives the call, hangs up, and calls the user's terminal back only if its phone number matches the account and password.

Encryption, the coding of stored data, can be costly, but it is technically straightforward. The federal government requires that certain personal and financial data, as well as national security information, be encrypted, but the policy has been implemented unevenly. In addition, major corporations also encode sensitive data. No code is unbreakable, particularly with the enormous, inexpensive computer power now available to potential snoopers, but encryption can discourage unsanctioned file use by making it more costly, less convenient, and slower.

Ironically, one federal agency is responsible for actually limiting the level of encryption generally applied to confidential civilian files. The National Security Agency participated in the committee which established the National Bureau of Standards' Data Encryption Standard in 1977. Reportedly, when IBM's representatives on the NBS committee argued for a 128-bit key, NSA argued successfully for a 56-bit key, which is much simpler to decode. NSA, it appears, wanted to keep the code simple enough so that an agency like itself with enough computing power could break it if it really needed to.

Operators of data banks can also take steps to restrict the unsanctioned use of computer records by authorized system users. The computer system at Hercules, Inc., a major chemical manufacturer, limits the functions of its terminals. *Business Week* reports that the company's accounting staff cannot look at personnel records, even though accounting and personnel departments routinely log on to the same machine.

Another, more effective way to limit the unauthorized use of personal information is to restrict the amount and kind of information stored in a file in the first place. Under the urging of the Business Roundtable and the U.S. Commerce Department, a number of large firms, including United Technologies, General Tire & Rubber, Chase Manhattan, and Cummins Engine, have policies limiting the kinds of data kept on employees and restricting who has access to such data. Northrop, a large military contractor that retains a great deal of personal data in order to obtain security clearances for its employees, keeps those records separate from its regular personnel files.

In fact, the technology exists to record each time anyone looks up a computer file, as well as the source of the connection, even if no change is made in that file. Users can be identified by password and/or by telephone number. For this purpose, computerized data banks offer an advantage over more primitive methods, such as hardcopy (paper) file indexes, since it is extremely difficult to keep track of card files or printed lists that can be used or changed by several people.

Operators of time-sharing services, which rely on "logs" to charge their customers, have been tracking systems users for years. Many accounting systems automatically note whoever has ordered a financial transaction. But operators of data banks containing personal information have not developed such procedures, for they have little to lose if the information gets into the wrong hands. Their business suffers only when their inadequate security is publicized.

Recording who has searched or modified a computerized personal file will not, in itself, halt unsanctioned use of the data. But if those people whose lives are described in the data bank are given the right to inspect that record, just as they may now examine their FBI files and credit reports, then "investigators" who abuse their privacy will face a significant risk of discovery. If Congress or the states pass legislation providing the victims of an illegitimate investigation with clear legal remedies, many would-be snoopers may decide that the information gleaned through their electronic searches is not worth the risk.

Sweden's data bank law, enacted in 1973, is a working model of legislation designed to limit the abuse of electronic file-keeping systems. The law established a Data Inspection Board with the power to regulate every public or private data bank in the country that keeps personal data. Residents are empowered to see all of their files; they can order corrections; and they can sue for damages. No data bank may keep criminal records, data on psychiatric treatment, political views, or even religion without permission of the Board.

■ The organized abuse of civil rights—eavesdropping, file-keeping, and file-matching—by public and private investigative agencies is extremely difficult to control. While the courts sometimes refuse to hear evidence collected by illegal means, even monitoring the behavior of highly secretive police and intelligence agencies is virtually impossible unless they offer their investigative results in court. Equally important, many high-tech surveillance methods that run counter to the purpose of the Fourth Amendment are presently considered lawful.

When legislative bodies do pass laws to protect privacy and civil rights, such as the federal Privacy Act of 1974, new technologies make possible surveillance techniques not fully treated by the law. Civil libertarians and high-tech snoopers are in a race, and the snoopers appear to be winning. Charles Marson, a law professor at Stanford University formerly on the staff of the ACLU, says that the more sophisticated the technology, the less likely it is to be covered by existing laws. "Why break into a house when you have the capability of remotely turning a telephone into a microphone?" he asks.

Federal law apparently allows private parties, public officials, and even foreign spies to intercept nonaural messages that many of us consider private, including computer communications, telegrams, teletype, facsimile or television transmissions, and telephone dialing data. In fact, the federal General Accounting Office suggests that voice transmissions that are coded into nonaural digital pulses for relaying over the phone network may also be subject to legal eavesdropping.

The Foreign Intelligence Surveillance Act of 1978 requires that government agents obtain a court order to tap certain types of nonaural communications only if the target has a "reasonable expectation of privacy." Implementing that rule, federal courts have ruled that law enforcement agencies do not need a warrant to track the numbers dialed on a person's phone because he knows that such information is already collected by phone companies. But the average American does not expect his or her dialing habits to be monitored. And even if everyone were aware that investigators could track their phone use, such surveillance would still be unreasonable.

Furthermore, international communications are not

fully protected by the Fourth Amendment. In November 1982, in a decision that stirred surprisingly little controversy, a federal appeals court ruled that the National Security Agency could lawfully intercept messages between U.S. citizens and people in foreign countries, even when there is no reason to believe that the Americans are foreign agents. In that case the government publicly admitted for the first time what critics and journalists had charged for years: NSA's surveillance equipment is used to acquire the overseas communications of individuals in the United States.

The case involved the government's surveillance of a Michigan-born lawyer, Abdeen Jabara, who for many years represented Arab-American citizens and alien residents in court. In 1967, the FBI began to investigate Jabara, and in November 1971, the Bureau asked the National Security Agency for information about the lawyer obtained through its monitoring of the international phone system. In response the NSA gave the FBI summaries of six overseas communications made by Jabara.

Current law requires that the FBI obtain a warrant from a federal judge if it wants to tap a phone conversation. If the FBI believes that the suspect is an agent of a foreign power, it can seek an order from a special secret panel of federal judges. Yet the court ruled that it can legally use information intercepted by the NSA without a warrant.

■ Until the mid-1970s, computer technology inherently centralized authority, giving the rich and powerful an effective aid in watching over the rest of us. Today, this remains the dominant trend, but individuals can now use microcomputer technology to monitor those who are watching them.

Presently, a patchwork of federal and state laws requires that the operators of many data banks, such as the FBI, credit bureaus, and large employers, allow people who are listed on their computers to inspect their files periodically. But it is difficult to determine which files are accessible, and the process of obtaining one's listings can be time-consuming.

With a microcomputer, however, the individual who wants to keep track of his or her files can develop a comprehensive list of file-keeping agencies. We foresee the development of commercially available computer programs designed to do precisely that, listing both public and private data banks as well as their rules of access. Again, while such reverse surveillance would not in itself halt the epidemic in privacy invasion, it might at least keep many file-keepers honest.

On a more political level, the widespread availability of inexpensive microcomputers has made it possible for dissenters of all political stripes to enlist high technology in their electoral, lobbying, and direct action campaigns. Virtually any organization, publication, or candidate can develop computerized lists of supporters and potential supporters, broken down by demographics, interests, or attitudes. For example, several national peace groups have developed a file of members and sympathizers, organized by congressional district, so that they can rapidly notify the constituents of "swing" Congress members when a key vote is scheduled. Such lists can be misused, or even stolen, but on balance they may provide an effective counterweight to the manipulation of personal data by credit bureaus, law enforcement and intelligence agencies, and mass marketers.

Random computer break-ins by computer hobbyists actually may counter the authoritarian use of computers. Hobbyists may in fact be the Robin Hoods of the information age. Those corporations, institutions, and government agencies that might wish to employ their computers to conduct blatantly illegal surveillance must think twice, because it is always possible that some teenager or college student might sneak into the system and stumble upon the transgression.

Similarly, computer-wise insiders with a commitment to the Bill of Rights could be a key line of defense against the eventual establishment of totalitarian rule. If some day the FBI or Secret Service decides to pick up those of us who are on its current version of the early 1970s Administrative Index, it would take only one savvy agent who values our tradition of individual freedom to crash that organization's computer system or to garble its instructions.

America cannot afford, of course, to rely upon conscience-stricken law enforcement agents or adventuresome hobbyists to protect it from the net of electronic surveillance that threatens to ensnare us all. To protect our society from becoming a police state—in fact, to avoid the self-censorship that accompanies the very fear of repression—new methods of invading privacy must be regulated as they emerge. Electronic data banks, be they law enforcement files or credit records, must be open to the people listed therein.

However, the greatest protection against instruments of surveillance has little to do with the particulars of the technology. We have to create a political environment in which a sophisticated vigilance against the suppression of independent thought and dissent is part of everyone's education and awareness.

Chapter Four
THE LEARNING GAME

■ ■ ■ ■ ■

California Congressman Fortney Stark (D-Contra Costa County) made his name in politics during the Vietnam War, when he displayed a huge peace symbol from the wall of the bank that he owned in Walnut Creek. Not too long ago, on a cross-country plane trip, he met another Californian who has managed to blend liberalism and money, Apple Computer founder Steve Jobs. A few phone calls later, Stark was sponsoring legislation to give Apple and other computer companies new tax breaks for donating personal computers to the nation's schools.

Though the legislation was never enacted, Stark has enthusiastically promoted educational computing. During a visit to San Francisco in early 1982, the story goes that Stark was demonstrating how anyone can use an Apple Computer to learn geography. He asked the student at the keyboard to type in "San Francisco" for a brief description of the city. But the student, even after several attempts, couldn't spell San Francisco.

Parents, educators, and politicians are all looking for easy ways to restore the effectiveness and reputation of America's school system. The personal computer, the most familiar instrument of the microelectronics era, seems to fit the bill. As a classroom aid, it is cheap, flexible, and easy to use. In 1981, less than one fifth of the nation's schools had at least one microcomputer; by 1983, over two thirds were using them.

In the words of U.S. Education Secretary Terrel Bell, the push to bring computers into the schools has become "almost a fad." Basic skills are frequently neglected, and the teaching potential of the new machines is also underutilized. Without a massive investment in software development and teacher training, microcomputers will bring little more than game-playing into the typical classroom. More important, the schools are failing to offer children a chance to learn *about* high technology and its impact on their lives.

■ Most children and teenagers like interacting with computers. The response is immediate. Positive feedback, like a high video game score, is rewarding. In many situations, mistakes are not bared before classmates. And, as in professional use, computers save time and worry. An English teacher at a private Connecticut high school told *InfoWorld* how his students responded to computers: "Word processing eliminates students' fears of poor spelling, poor handwriting. They'll take all kinds of risks in their writing with word processing, and they're much more inclined to do revisional work. I am seeing a willingness to experiment and play around with the language."

Today, classroom computers help teach virtually every subject, from algebra to zoology. Some of the best educational routines actually combine the excitement of game playing with specific learning tasks. For instance, students can learn typing by quickly keying the words displayed on "spaceships" or "missiles." Or they can learn note taking and problem solving in role-playing adventure games designed to teach those skills.

Still, some critics argue that microcomputers are too sophisticated for most classrooms. The Committee on Basic Skills Education, representing the producers and advocates of older, programmed-instruction equipment, argues that "there are simpler, less costly and more efficient" teaching aids available for elementary education. Committee spokesman Eric Burtis, who is the president of Centurion Industries in Menlo Park, California, says: "What we're doing with general purpose computers is complicating the learning process with highly sophisticated, expensive hardware systems. We believe those funds should instead be used to improve curriculum and strengthen basic academic skill education, using established, simplified techniques and technologies."

Burtis's claim that funds are being misspent rings true. But the real problem with the use of microcomputers in the schools is that all too often they are being used like Centurion's "simpler, less costly" systems. When used to their potential, microcomputers are a powerful teaching resource, offering immense flexibility to the teaching process. Frequently, however, they merely emulate the programmed instruction or rote learning of earlier generations of teaching machines.

In the absence of creative software, computers are little more than electronic flash cards or page turners. Programmed instructional materials—whether presented on paper, with primitive teaching machines, or on microcomputer screens—attempt to mimic that process by regulating the rate at which students are presented with new material. But anyone who has used the existing programmed instruction texts knows how crude and inflexible such systems are. Frequently, they merely provide some students with busy-work so that teachers can focus their energies on others.

The chief obstacle to the development of adequate educational programming is a financial one. Entrepreneurs are reluctant to go into educational software publishing because users can easily duplicate the programs without paying. The nation's poverty-stricken school systems are readymade networks for the "informal" distribution of software. So software companies are likely to design educational programs for the more lucrative home market. Some of these home-oriented packages can easily be used in the schools; however, most do not achieve the standards sought by experienced educators.

To fill the software gap, school systems are beginning to review, distribute, and even to write their own programs. In California, for example, the San Mateo County Office of Education has joined with a group called Computer-Using Educators to form Soft-swap, which helps circulate programs already in the public domain. And the Philadelphia school district has developed its own computer-assisted instruction materials.

The Minnesota education department has established what many consider a model program for bringing computers into the public schools. Educators there have compiled a list of "high-quality" programs that school districts can purchase with state aid. Minnesota has actually paid for the development of dozens of new programs. For a one-time payment, other state school systems can make an unlimited number of copies.

Minnesota's approach offers the greatest hope for the preparation of high-quality educational software. Already individual schoolteachers across the country have found it necessary to write their own programs, without compensation, for classroom use. If public agencies support such efforts through financial help and technical assistance, as well as by providing nonprofit distribution channels, they can establish a massive library of creative, teacher-tested software. Such an effort will require careful judgment, however, to ensure that centralized funding authorities do not impose uniformity and mediocrity upon the school districts receiving their support.

Furthermore, even with financial support, it will be years before educational software simulating the teaching methods of top-notch teachers is available. Good instructors learn the abilities, problems, and idiosyncrasies of each of their students. To the extent that they can afford to give individual attention, these educators tailor their instructions to each student's needs. This is an interactive process, requiring constant adjustment. Today's microcomputers are both cheap enough and capable enough to augment human teachers by providing individualized, interactive teaching programs. Unfortunately, it appears that it will be some time before programmers and teachers get together to design learning routines complex enough to do this job well.

The development of new educational software is essential to the future of the computerized classroom; at the same time it may divert scarce resources from the teaching of fundamental skills, the enhancement of traditional teaching methods, and the use of alternate technologies. Many teachers complain that they must buy simple classroom supplies from their own personal funds, because school budgets are increasingly going for software and other computer-related expenses. Videotaped teaching aids, often more practical than computers, are frequently neglected. In some classrooms, computers supplant, rather than supplement and support, the teaching of reading, writing, and arithmetic.

■ With enthusiasm reminiscent of the national science education effort launched following the Soviet Union's 1957 orbiting of *Sputnik,* America has embarked upon a national campaign for "computer literacy." TV and radio commercials warn that any youngster denied his or her own computer will be permanently handicapped educationally, socially, and economically. Parents, when they visit their children's schools, seem more interested in watching the school computer in action than in talking to the teachers. Many educators have jumped on the bandwagon, too. The College Board, for instance, concludes that "Competency in [computer] use is emerging as a basic skill complementary to other competencies, such as reading, mathematics, and reasoning."

The proponents of computer literacy training overstate its importance. It is neither essential nor fundamental. In coming years, few people, even those who use computers on the job, will need to know how to program a computer. Manufacturers are working hard to develop "user-friendly" machines—computers that can be operated by a beginner. Thus, many students today are being taught programming languages and techniques that may soon be obsolete.

Still, computer literacy is a valuable skill. In the long run, it may open the door for a job in high-tech industry. In the short run, computer training can help a youngster learn how to reason and communicate.

In good computer literacy training, the student learns to command the machine, unlike drill-and-practice, where the computer is in control. Properly designed, computer literacy courses give students a taste of the many things a computer can do. They can acquire the essentials of word processing, or they can discover how to sort a mailing list with a "data base" program. And they can learn the elements of logic.

Twenty years ago, artificial intelligence researchers at MIT invented Logo, a computer language which offers the best in computer literacy training. Logo allows children to program simply by manipulating a turtle symbol on the video screen. To draw a twenty-unit line on the display, the student merely keys in "Forward 20." To form a corner, he or she types "Right 90" and "Forward 20." The turtle dutifully responds, and if instructed, it remembers the sequence of commands. The novice programmer has thus created a corner-making subroutine, to be reused drawing squares and other geometric figures.

Logo developer Seymour Papert, who is a proponent of the theories of the Swiss psychologist Jean Piaget, originally envisioned Logo as a free-form tool that would allow children to understand abstract conceptual ideas in mathematics and programming. With Logo, a student does not merely answer questions provided by a teaching machine; he creates his own learning path.

When taught by well-trained teachers, Logo lives up to Papert's objectives. In practice, however, Logo is frequently used in classrooms as merely one more form of mechanization to permit teachers to carry a larger class-load. Applied indiscriminately, Logo will go the way of the new math. This would be a tragedy, for Logo as an exploratory environment for young children can convey both the power and the utility of personal computers.

Furthermore, computer literacy is all too frequently treated as a high-tech form of vocational training. Political and business leaders see computer literacy as a requirement for reestablishing America's industrial competitiveness in the world marketplace, and they are pushing the concept upon America's youth. For instance, David Kearns, president of Xerox, says: "Electronic literacy—or illiteracy—is a concept that came out of my industry, and it means quite simply the ability to use a computer. That ability will be necessary for most jobs in the future."

However, the notion that computer literacy is rapidly becoming an economic necessity is absurd. Learning to program may help give some youngsters the opportunity to move eventually into high-status, well-paying programming jobs, and many will need computer skills later in their education. But both computers and the nature of computer skills are changing rapidly. In the future the division will be extreme; professional jobs will require a general competence in verbal communication and mathematical reasoning, while most jobs will require little or no skill at all.

Although high-tech occupations are growing rapidly, the five vocations expected to generate the largest *number* of new jobs are all in positions requiring little training and no computer literacy: janitors, nurses' aides, sales clerks, cashiers, and waiters and waitresses. During the twelve-year period from 1978 to 1990, 150,000 new jobs for computer programmers are expected to open up; for kitchen helpers and fast-food workers, there should be 800,000 new positions.

Many low-tech workers will tend microprocessor-controlled machines, such as cash registers, warehouse terminals, and robots, but computer training will be of little value to them on the job. Computer designers are constantly making their equipment easier to learn and use; and employers are in fact using high technology to reduce the skill demanded of their employees. For example, cashiers at many supermarkets no longer need to remember prices, since their check-out stands are equipped with laser bar-code readers; counter workers at hamburger stands need only press a key marked with either the description or a picture of a sale item. And most cash registers automatically calculate change.

■ Today's students need to learn about the many ways in which they will be interacting with computers and other high-tech equipment in the years to come, but they do not necessarily need programming classes. Rather, young people—as well as adults—need general instruction on the capabilities and shortcomings, uses and misuses of computers, if they are to act responsibly as citizens, workers, and consumers in the information age.

If America's schools carefully assess the role of computer education, computer citizenship training will soon be the central feature of the computer curriculum. Just as many school systems emphasize "driver's education" over behind-the-wheel "driver's training," "computer literacy" may be eclipsed by "computer citizenship." Unlike driver's education, however, computer citizenship will be graded into units appropriate to each age and ability group.

At present, very few schools offer courses in computer citizenship, but a handful of leading educators are beginning to speak out about its importance. Ernest Boyer, who prepared the Carnegie Foundation's 1983 critique of high school education, suggests that computers can be valuable tools, particularly for remedial learning, but stresses that the first priority is a broader understanding:

> In the future, computers for the nonspecialist will be so convenient that little technical skill will be required. Few citizens of tomorrow will spend time at a computer keyboard or will write a computer program. Therefore, the first goal should not be hands-on experience for students. Rather, the urgent need is for students to learn about the social impact technology has played and will play in their lives.

Computer citizenship training will explain, with hands-on demonstrations, exactly what is meant by computer applications such as electronic banking, graphics design, and computerized inventory control. However, students will not master such applications beyond what is needed for their classwork—simple word processing, perhaps. By the time they are ready to use their skills in the workplace, the techniques will have changed.

Ideally, computer citizenship coursework will focus on some of the questions raised in this book. Does electronic banking threaten personal privacy? Will word processing downgrade or upgrade the skills and status of office workers? Will computerized factories threaten our sources of income? Of course, third graders won't be able to answer these questions, but high school students can learn to ask them.

As they mature, students will develop their views about high technology. What role should computers play in society? To what degree should they as individuals choose to work with computers and other forms of modern technology? And how can computers be used to open up creative, rewarding jobs for the typical American worker?

Preparing young people to be effective members of the information society depends, of course, on more than specialized training with and about computers. The same basic analytic and communicative skills that will enable today's children to adapt to the workplace of the future are even more important if they are to deal with the political and cultural changes already being wrought by the application of high-technology electronics.

Despite the ineffective way in which computers are currently being used in the home and in the schools, universally available computer education will help establish equity in opportunity. Though not everyone needs to learn how to program a computer, every child deserves the chance to become part of the fraction of the U.S. workforce that relies on high-tech skills.

Since the early 1970s when Atari placed Pong, the original video game, on the market, most American youngsters have had at least limited access to computers. After all, video games are merely computers that use knobs or joysticks for input. Today's games are much more exciting and colorful, and for a quarter or fifty cents, anyone can play.

Yet, while access to computer technology may be universal, it is by no means equal. Working-class kids play video games at home or at arcades; the more affluent youngsters have their own home computers. Chances are that most rich kids use their computers for game playing most of the time, but their keyboard-oriented machines at least offer the chance to learn. Poor children may become wizards at Space Invaders or Pac-Man without ever having a chance to learn how to program.

Senator Frank Lautenberg, founder of ADP, one of the

country's largest data-processing service firms, warns: "The concept of computer literacy defines a new type of illiteracy and the potential for new and distressing divisions in our society." Citing figures on the explosion in home computer ownership, Lautenberg adds: "Those computers are being acquired by the affluent, reinforcing the disparities in opportunity."

Overcoming those disparities and establishing equity is the best justification for putting computers into the schools. Although rich school districts have more computers, more technically trained teachers, and stronger math-science courses than poor districts, computers in poor districts may give poor students an exposure to computers that they can't get at home.

Grant programs like Apple Computer's that provide computers, software, or training to *all* schools can help to equalize the access to computers, but targeted programs are even more effective. One such program is a joint project of the Xerox Corporation and the National Urban League, whose chairman is Xerox president David Kearns. Xerox is providing one thousand personal computers, valued at $5 million, to the Urban League. A committee representing Xerox, the Urban League, and the educational community has been formed to select the recipient schools. The machines are being targeted specifically for urban schools with large minority populations and financial need.

Setting inner-city kids in front of video screens will not, of course, guarantee that they will receive the computer literacy training now taken for granted in many affluent suburban schools. In computers, just as in any other subject, schools need the funds to attract, retain, and support competent teachers. They need software written to appeal to the cultural interests of racial and ethnic minorities, since in many cases existing home-oriented software was designed for the white middle class. And they need a commitment to all forms of computer training, not just remedial drill-and-practice or vocational training.

■ Unlike racial minorities, young white females from both moderate-income and affluent families can gain full access to computers. Yet our culture still discourages them from doing much with them. Few girls either have or use a computer at home, and girls sign up less often than boys for computer courses. Stanford researchers Irene Miura and Robert Hess found that three times as many boys as girls enrolled in computer camps and classes at all age levels. The proportion of girls in beginning and intermediate classes was 27 percent; that share dropped to 14 percent in advanced programming classes, and to 5 percent in higher-level courses teaching assembly language.

Miura, Hess, and others who have studied the computer literacy gender gap have found that parents encourage boys to use computers more often than girls. Mothers, as role models, tend to use computers less than fathers. In many cases girls shy away from computers because the machines are associated with action or shoot-'em-up video games, which better match the socialization of American boys.

The tracking of girls away from computers mirrors the traditional tracking of girls away from math and science, but it is nevertheless ironic. Because programming is a verbal skill as much as a mathematical skill, many women who as girls were encouraged to avoid math are now working as computer programmers. In 1982, 3.6 percent of the engineers in the United States were women; but 28.3 percent of the computer specialists were women.

Overcoming the computer gender gap should be

easier than overcoming the income gap. Parents and teachers, as well as peers, can reinforce the use and study of computers. Women in computing careers can meet with students to discuss the benefits and disadvantages of working with computers. And as long as girls as a group pursue different interests from those of boys, software can be written to appeal to those interests.

Computerized learning has already entered the lives of our children so rapidly that few parents, teachers, or public officials have been able to assess, let alone channel the technology. In fact, educators did not ask for high-tech teaching tools. Computers were foisted onto the nation's school system by the demands of the competitive marketplace, long before the schools were ready.

To most manufacturers, the schools are primarily a vehicle for building a consumer market for personal computers. Though school sales are significant—in 1982, for instance, Apple sold one quarter of its volume to educational institutions—the home market is five times as large, and it is much more profitable. In school, youngsters learn to use and develop a taste for personal computers. Home computers not only cater to that interest, but they help students to get ahead in computerized coursework. The pressure on parents to buy is enormous.

In fact, computer manufacturers consider placing their machines in classrooms so useful that they give them away. Apple, for example, provided ten thousand Apple II's to schools in California, and it promised a computer to virtually every school in the country if Congressman Stark's tax proposal passed. Though Apple's top executives believe strongly that their machines are good for kids, their motivation is primarily economic. If the parents of just a few students each year in every

Apple classroom buy an Apple because that is the machine their sons and daughters know, the company will ring up a tremendous profit. Accordingly, Apple ads tout its machines as the computers found in the most classrooms.

Among the producers and advocates of educational computer equipment, there are many who envision a fully electronic classroom, in which each student has a microcomputer or terminal permanently fixed to the desk. Each machine is tied to a network controlled by the "teacher," who is little more than a computer operator providing appropriate prepackaged learning routines and games. The teacher can also surreptitiously peek at the work of any student at any time merely by pressing a few keys on the master keyboard. Such classrooms already exist, but fortunately they are rare. Should educators ever adopt the automated classroom, computer companies would rake in revenues that will dwarf the current educational computing business.

■ With all the attention being paid to the educational value of home computers, it is easy to forget that they are still primarily a recreational medium. One computer executive reportedly admitted, "Education is the name of the game for parents. The name of the game for kids is games." The bubble has burst in the market for video game hardware and cartridges, but computer games, whether played on personal computers, video consoles, or arcade machines, remain extremely popular.

Electronic games are popular because they offer young people—of all ages—undivided attention and immediate feedback. Their simulations, approaching science fiction fantasy, give users a sense of power. One can be a successful game player even if one is not pretty, tall, popular, or academically successful. In many video and computer games, teenagers can easily outclass the adults who dominate their lives. Most important, the games are fun.

In fact, adults sometimes fear high-tech games precisely because they are so popular. Many teenagers appear addicted to the beeping screens, but there is no evidence of widespread psychological dependence. Parents are often concerned that their children spend too much time playing video games, but excessive game playing is not much different from too much television, music, or sports. The solution usually lies in the offering of enjoyable alternatives.

Electronic games are not, as some critics suggest, mindless. Most rely heavily upon quick reactions and good eye-hand coordination, but top players usually learn to recognize patterns and develop strategies to maximize their scores. Those parents who worry that Pac-Man does not sufficiently challenge their child's mind can always offer an electronic chess set or some other thought-based game.

Although many of the parents and citizens who have rallied against the rise of video parlors border on the paranoid, the way in which electronic games trivialize violence is a genuine cause for concern. Youngsters don't go home and beat up their little brothers or sisters because they have just closed the arcade, but the continued exposure of large numbers of people to electronic war games will make them much more tolerant of organized high-tech violence—in other words, modern warfare. Anyone who routinely destroys planets without ever seeing a drop of blood is a prime recruit for America's new legions of electronic warriors.

Surgeon General C. Everett Koop found himself in the hot seat without a joystick in November 1982 when he speculated publicly that video games fostered violence, or at least made children more tolerant of violent behavior. Koop didn't cite any scientific studies, but he predicted that research would eventually bear him out. However, demonstrating the causes of violence may be even more difficult than linking cigarettes to cancer, since, in the words of a black leader of the 1960s, "Violence is as American as apple pie."

The controversy is not new. Americans have debated cartoons, television, boxing, football, and other forms of entertainment for decades. Those who oppose one or more of those activities suggest that violence on the screen, on the field, or in the ring begets violence in the home and on the streets. Defenders argue that, on the contrary, sports and entertainment help release tensions that otherwise would lead to violence.

Games, in this regard, may pose less of a problem than network television. The argument that TV viewers become inured to violent behavior makes sense. On TV a child sees a constant chain of killing and other forms of violence; and while the scenes may not be as gruesome as in R-rated movies, the victims are identifiable human beings. In video games, however, the victims are depersonalized blips on the screen. Blasting the electronic image of a spaceship into hundreds of pieces does little more than knocking over ten bowling pins to accustom the player to antisocial violence.

Unfortunately, the depersonalized mayhem of the typical video game resembles a more serious form of violence, high-tech warfare. One of the major reasons that the U.S. military is working hard to automate its future battlefields is to overcome the reluctance of soldiers to fire on targets that might include innocent civilians. Electronic targeting techniques make the victims faceless. The Marine of the future who destroys a village or the ground controller who nukes Petropavlovsk may rest comfortably, never having seen blood and guts. They will simply be playing grown-up video games, in which the results appear just as they do at the arcade, with blips on the screen and kill totals on the digital monitor.

The games may actually be a little more sophisticated than the real weapons, but they are still preparing young Americans for the electronic battlefield, and the military knows it. For instance, Julie Reed, the Navy's top recruiter in 1981, found a number of her enlistees at Northern California video arcades. She told the *Associated Press,* "I just ask them if they know the Navy has sonar, radar and computer weapons that work just like the games. That interests them." In fact, game players may find the "real thing" less exciting, because few soldiers, sailors, or pilots see as many targets in their entire military careers as the average teenager vaporizes in one game. Nevertheless, a Navy medical researcher suggests, "the ideal pilot may someday be a middle-aged, short guy who wears glasses, has a large bladder, handles stress well and spent his youth playing video games."

Video war games don't provide any useful military training, since the average electronic warrior is little more than a button-pusher. Rather, the games are habituating their devotees to high-tech weapons and generating enthusiasm for future wars. Working-class youths who accept the offers of recruiters like Julie Reed may avoid dead-end jobs in data entry, robot operation, or fast-food sales. But without fully recognizing it, they will have enlisted in jobs that have even more deadly ends.

There is a need to develop exciting but less violent alternatives to video war games. Emerging technologies, such as lifelike computer graphics and speech synthesis, are creating new possibilities in electronic entertainment. Those technologies can open new vistas in music or art; and they can bring a wide range of sophisticated role-playing games or challenging electronic puzzles. But unless the market is there, the designers and manufacturers will simply add more bells and whistles to the same old war games.

■ Whether used for word games, Space Invaders, or advanced programming, computer and video games are definitely influencing the ways in which young people relate to each other. On the one hand, the anonymity of interacting with computers may help shy children to overcome their fears of human interaction. On the other, "addiction" to the rewards of video technology may isolate young people, reducing their opportunities to develop simple social skills.

Video games and personal computers earn the attention of young people by promising immediate, generally positive feedback. Used in moderation, they can help unpopular or shy children to develop self-confidence. The machine does not care if the person pushing the button has pimples, stutters, or is overweight. In the arcade, the unattractive child can win the admiration of his or her peers by developing game-playing skills.

Even in the classroom, students can slowly build their confidence by telling their answers privately to a computer. Many teachers report that shy students frequently communicate better with both students and teachers after they have worked with a machine. An eighth-grade math teacher in San Francisco who didn't have enough Atari computers to go around in his multicultural class says that the shortage forced his students to share machines and work together as they never had before.

Games also can be organized to promote interaction. Many games, of course, are designed so that they can be played by two people. But there are more sophisticated computer games, such as the role-playing adventure games, that require the interaction of several players. In one, players act as the "Star Trek" crew, solving interstellar problems in the style of Kirk, Spock, and their companions. At present, complex multi-role games are available only to groups with networked computer terminals. If interest grows, however, they may soon be packaged for the general consumer market.

Still, a child who relies entirely on video games for entertainment or spends every spare hour studying programming may never learn to interact with other people. Team sports, school dances, Scout trips, and the like may at first be more difficult for some, but in the long run such activities are necessary if a child is to function with confidence in the adult world. But the excitement of video games or computers cannot be blamed when young people eschew such activities. The problem lies in the lack of encouragement, help, and attractive alternatives.

■ High technology, whether it is used for education or entertainment, is preparing young people for their adult lives. Unfortunately, the status quo in the schools, the home, and the arcade reinforces the status quo in adult society: inequities in our socioeconomic structure and militarism in our foreign policy. Offering computer literacy and citizenship training to all young people, regardless of background, and disarming video games may help to prepare young people for a better world, but the preparation will be useless if they are not given the tools with which to build that world.

Even without reforms in high-tech education, the technology is already creating rising expectations that adults as individuals and society as a whole may have great difficulty satisfying. The child who learns to program at age eight may not relish working in data entry.

The teen-age conquerors of galaxies may not be satisfied when they come down to earth and find themselves lasering auto parts or feeding chips to robots.

Our society, like that of each generation before, must change to accommodate the goals, if not the fantasies, of today's youth. Many of those goals—freedom of expression, access to knowledge, equal opportunity, the peaceful resolution of international disputes—have been around for centuries. But high technology thrusts upon us, at unprecedented speed, complex decisions as to how we can achieve them.

Chapter Five AUTOMATION NATION

■ ■ ■ ■ ■

In Fremont, California, on the outskirts of Silicon Valley, stands a vast factory that is two football fields long, one football field wide, about three stories high. Inside, fewer than three hundred Apple Computer employees (less than one hundred production workers per shift) assemble Macintosh personal computers. Utilizing the latest in factory automation systems, the Macintosh plant is designed to produce half a million machines each year. Yet direct labor, so the company reports, accounts for less than 1 percent of production costs.

Production employees request parts by keying in simple instructions at their own computer terminals. Battery-powered robots, controlled by a microprocessor-based traffic management system, respond. Other machines automatically insert 85 percent of all the Macintosh components into printed circuit boards. A completed Macintosh runs a series of tests on each assembled circuit board. Following final assembly, the new machines are placed on "burn-in" racks for additional testing. Each seven-tier rack is controlled by a single Macintosh microcomputer.

In a sense, Apple's Macintosh plant is a science fiction writer's dream. State-of-the-art computers are reproducing themselves for mass consumption. It will be many years before such machinery is fully capable of "reproduction," but the Macintosh factory poses a clear question: Today, in economic terms, what are people good for?

■ The displacement of men and women by machines is nothing new, yet both the scale and the pace of the current round of automation, based upon the microprocessor, are without precedent. In the past, most automatic production equipment, whether powered by water, steam, electricity, or gasoline, could carry out only a single, repetitive task. Where flexibility was required, machines were directed by workers.

However, systems based upon the microprocessor—a computer on a single silicon chip—can adjust themselves or be reprogrammed without constant human guidance. Flexible automation is cheaper than the older forms of mechanization, for the equipment need not be replaced or retooled every time a product line or a process is changed.

Moreover, the cost of chips, the building blocks of all high-tech machinery, continues to fall at an exponential rate. While other industries find their markets limited by high or growing manufacturing costs, the producers of "labor-saving" equipment are constantly cutting costs and prices.

Since high-tech machinery is so cheap and flexible, automation is spreading to portions of the economy untouched by previous waves of industrial mechanization. Computerized equipment is used by workers in almost every occupation, from accountants to zoologists. Microprocessors are ideal tools for information-based activity, such as banking, publishing, and corporate management: by calculating, sorting lists, retrieving data, and editing documents, computers do the work of the mind, just as the steam-driven loom mechanized the motions of the body two centuries ago. And computer scientists are now developing so-called expert systems —machines with elaborate software designed to replace skilled professionals, such as doctors, geologists, and even computer scientists.

Of course, automated information processing is reshaping the modern factory. In the past, the knowledge, data, and skills necessary to operate an industrial plant rested in the minds of managers and skilled workers; today, manufacturers use computerized systems to control the flow of materials and parts and to measure the output of workers. They rely on microprocessors to operate machine tools, painting equipment, valves, ovens, and virtually every other type of production equipment.

Microprocessors can be used to help automate virtually any work in which information plays a role, even in the most unusual workplaces. For example, a recent *Wall Street Journal* ad describes one of Intel's latest industrial automation systems: " . . . the moment the cow ambles into the feeding station, a tiny transducer attached to her collar identifies and announces her arrival to a system based on an Intel microchip. In the flick of a tail, the system selects and doles out the proper quantity and exact type of feed best suited to this particular cow." Intel's cow illustrates the flexibility of microprocessor technology and its ability to penetrate virtually all economic activity. Today's computerized systems can alter their functions based upon sensors which identify parts, measure thickness, or recognize cows.

Finally, the continuing process of squeezing increasingly complex integrated circuits onto a tiny flake of silicon—without increasing the cost of the chip—permits another cost-saving form of automation, the design of products requiring fewer components. For example, Apple's Macintosh, designed to minimize its cost of production, contains only fifty integrated circuits, 75 percent fewer than IBM's roughly comparable PC. Obviously, it is cheaper to skip stuffing extra chips into printed circuit boards than it is to use even the most efficient automated insertion system.

Automation undoubtedly increases productivity and makes money for the company, but it frequently hurts individual workers and burdens society as a whole. The computerization of the American workplace is well on the way to displacing millions of workers, forcing them either to relocate or to retrain, and in many cases to accept a significant loss in status and income. No one really knows how many jobs will be lost or gained as the chip goes to work. There is no reason to believe, however, that high technology on its own will solve the problem of unemployment created by workplace automation.

Before long, just about everyone's job will be transformed by the introduction of computerized systems. The apparent uniformity of high-tech work, in which people in widely varied occupations all sit or stand at similar video display terminals, masks the growing polarization of the American workforce. If present trends continue, a large minority will find their jobs more challenging and better paid. The majority, however, will be trapped in low-skill, low-wage positions. And throughout the economy, computers will give managers tighter centralized control over the workplace, with the ability to monitor closely and pace the activity of individual workers.

Not surprisingly, those who are hurt by automation may have the impulse, like the followers of Ned Ludd in England two centuries ago, to take a hatchet to the machines. But there are far more effective responses. Some American labor unions, following their concerned European counterparts, are beginning to demand a say in the latest round of automation. Through labor contracts or government rules, workers can seek shorter work weeks in exchange for more jobs. Tax and investment policies, now used to promote economic growth, can be designed to maximize employment.

- Automation is a slow process, since there are both economic and social obstacles. In other words, the fact that it is technically possible to mechanize a job does not mean it will necessarily happen. While it is difficult to think of a clerical or factory occupation that could not now be automated to one degree or another, the fear that we might all be thrown out of work tomorrow is unfounded.

For example, one would expect the publications which cover computer technology to sport offices incorporating the latest in high-tech equipment. But the writers at *MIS Week,* a paper written for and about managers of corporate data-processing systems, did not get their first video terminals until early 1984. At that time, journalists at *Electronic News,* a leading high-tech trade publication also based in Fairchild Publications' Manhattan office building, were still using manual typewriters.

To the average person, labor unions might appear to be the most serious roadblock to automation. After all, railroad unions have for decades insisted on extra, "featherbedded" personnel, and newspaper unions have fought tooth and nail to protect the jobs of their members against computerized typesetting. Yet the overall record shows the opposite. Unionized industries have been among the first to apply computer technology. Companies which pay high union-scale wages have a greater incentive to invest in labor-saving machinery, and many unions have for decades accepted the "productivity bargain." In exchange for increased wages and benefits, they have agreed to raise output per person-hour, even when that meant fewer jobs. Strong unions, such as the auto workers, machinists, and communications workers, are beginning to bargain over technology-related issues, but they have not opposed automation per se.

Managerial work has generally been automated slowly, despite the proven value of executive work stations. Corporate executives—particularly older, entrenched managers—often appear to be the most reluctant computer users, although many use computer-generated information. Work stations, in their view, are like typewriters, the tools of subordinates in outer offices and typing pools.

Even when there are no human obstacles, it takes time to redesign workplaces and train workers. And most important, automation is costly. Though it is likely ultimately to reduce the operating costs of a business, the initial investment in new equipment may be staggering. Apple Computer, when introducing the new Macintosh, needed to build a new plant in any case, so it could justify the installation of the latest factory automation systems. Firms that must abandon old equipment to automate face a much more difficult cost calculation and the tendency is to introduce new technologies more gradually. GM, for example, had to create an entirely new division in order to take full advantage of existing computerized manufacturing technology.

There is a further economic obstacle to automation. As workers face displacement by machines, their ability to bargain for decent wages deteriorates, and as wages slip, the advantages of automation decrease. It is not surprising, therefore, that the U.S. occupations projected for the greatest absolute increase are such low-paying jobs as fast-food cooks, hospital orderlies, and janitors. Much of this work could be automated, but it isn't worth it to the employers. Even the cheapest, most flexible machines still have trouble competing with the lowest-paid workers.

Nor does the application of state-of-the-art technology to the workplace automatically throw people out of work. Since dairy farmers presumably did not remix food for each cow before Intel introduced its sytem, the automated barn is displacing no one. Text-editing systems can be used to reduce the typing of multiple drafts of memos and articles, but in many cases the availability of word processors allows or even encourages an ongoing revision process, requiring even more labor. Some offices appear to have introduced a new wrinkle into Parkinson's Law: The volume of written materials expands to fill the capacity to type and edit it.

Thus, automation is more like a glacier than a flood. The process is massive and inexorable, yet it moves slowly. Human-made, it can be channeled and controlled; but it is still deadly to those who lie directly in its path.

Automation may be proceeding more slowly than the technology would permit; nevertheless, it is directly affecting the livelihood of millions of people. In the United States, as three Penn State labor scholars report,

"Researchers estimate that new technology will affect as many as 45 million jobs—about half of which are currently held by union workers. Of this number, approximately 25 million workers will be affected in the most drastic way—their jobs will be eliminated."

Technological unemployment affects more than the workers who lose their jobs to machines. If a large percentage of the American labor force is thrown out of work, who will support them? Who will buy the goods? Who will pay the taxes? Even if automated factories, farms, and service industries could produce all that we as a society can consume, we would need to develop new ways to distribute their goods and services.

Nobody really knows how many jobs are or will be lost to high-tech machinery. In the case of North American auto plants, for example, it is sometimes difficult to know whether to blame bad management, oil prices, design flaws, robots, or Japanese competition for the decreasing workforce. No one denies that the American economy is undergoing massive changes. Structural unemployment is substantial, and the impact on individual workers and communities is significant. But the long-term impact on the American economy as a whole has been subject to debate.

For the past two hundred years, critics of industrial technology have warned of massive, continuing unemployment, yet it hasn't happened. Most displaced workers have found new jobs. To be sure, a lower jobless rate would be better. Nevertheless, despite the warnings, capitalism not only in the United States but throughout the industrial world has demonstrated its ability to maintain a relatively high level of employment.

Yet there is cause for alarm. In the past, displaced agricultural workers could find jobs in industry. In this century, jobs lost in industry were replaced by jobs in the service sector. But today's flexible microprocessor technology is ubiquitous. Automation is occurring in virtually every industry. Other than high-technology industry itself—the production and servicing of electronic equipment and the development of software—it is difficult to imagine where long-term employment growth will occur. And high tech, though it is growing rapidly, is still not likely to absorb large numbers of the unemployed, particularly since microelectronics, computer, and communications firms are automating as fast as, if not faster than, other industries.

The low U.S. birth rate of the past decade may help to hold down unemployment. The U.S. labor force is growing at a much slower rate than during the 1960s, when the post–World War II boom babies entered the workforce. Some economists are pleased with this trend, because they point out that it will not be necessary to create as many jobs to maintain current levels of employment.

But they are shortsighted. Since working Americans pay the taxes which support the Social Security system, new workers will be saddled with an ever-increasing tax burden. After all, fewer wage earners will be paying into a fund that must support a growing number of retired people. Keeping older workers in the workforce longer, the most likely long-term solution to this second problem, reopens the original problem: There just may not be enough jobs to go around.

Historically, industrial countries overcame the problem of structural unemployment by finding or developing new markets. Much of the success of American capitalism can be attributed to its vast, growing domestic market. But an integral part of that industrial growth, not only for the United States but for Japan and Europe too, was the availability of nonindustrial markets. Through colonies and economic dependencies, burgeoning industries were able to market their surplus products.

Poor countries have historically paid for the import of manufactured goods by exporting primary agricultural goods and minerals. More recently, they have borrowed from rich countries to subsidize their imports. This trade has boosted employment in the industrial countries, but it has undermined the economic position of most Third World countries and their populations.

Leaders in many industrial societies, particularly in Europe, fear that the growth of high technology may turn their countries into economic backwaters as well. Recognizing that such high-technology equipment as computers, robots, and telephone switching systems is likely to displace large numbers of workers, governments are competing to attract the one manufacturing industry likely to benefit from economic restructuring: high-technology electronics. So competition for high tech is fierce. Governments use the "stick"—trade and procurement restrictions—and the "carrot"—research programs and even direct subsidies—in order to promote and recruit high-tech investment.

Those who are battling to become the next Silicon Valleys share an unstated assumption: High technology is restructuring the global economy so thoroughly that there will not be enough jobs to go around. Even the winners, the handful of countries that do manage to establish successful high-tech centers, will face uncertainty in the global market. Can the French, if they successfully develop a competitive, indigenous computer industry, expect to export machines to Italy and the Netherlands, if those two countries face a crisis of unemployment?

Nobel Prize–winning economist Wassily Leontief argues that a key factor in the ability of capitalism to absorb technological unemployment was the long-term reduction in the average work week, from sixty-seven hours in 1870 to forty-two hours at the end of World War II. He attributes the growth of unemployment since then to the relatively constant average work schedule. To overcome unemployment, he supports a shorter work week as well as income subsidies for the jobless. Leontief's program is practical. Most people would accept a slight drop in pay in exchange for shorter hours.

■ Every work morning an American Airlines jet arrives on the Caribbean island of Barbados and unloads a quarter-ton of used ticket coupons. The tickets are taken to the airline's Caribbean Data Services (CDS) subsidiary, where over two hundred women, each earning less than $3 an hour, key data from the coupons into a local computer system. Almost immediately, CDS transmits the information via satellite to the company's central data-processing operation in Tulsa, Oklahoma.

High technology, combined with advances in both surface and air transportation, has sparked seemingly contradictory trends in the location of industrial production and corporate administration. On the one hand, administration and communications work is being centralized; on the other, production, including routine office work, is being spread out across the country and in many cases, around the globe.

In fact, centralization of management and deconcentration of operations are two sides of the same coin. High tech enables companies to separate management from production geographically, allowing them to allocate each function to sites where the costs are lowest.

High-tech equipment could easily be used to decentralize authority in corporations and governmental institutions, but that is rarely the case. Now, with the ability immediately to transmit and receive data from remote facilities, executives at headquarters can perform functions previously carried out by local managers. Centralized processing centers offer large institutions a standardized data base from which to run their operations, as well as certain economies of scale.

This centralizing of operations is most visible in communications businesses, such as phone companies and telephone-based services. In the old days, operators, whether providing directory assistance or taking catalogue orders, had to be located in or near the areas they served. Today, telephone companies are shutting down their small branch offices and using high-tech switching equipment to feed calls, in order of receipt, from a wide geographic area to a bank of operators in a regional office. Catalogue sales companies have become even more centralized, replacing local units with "800-number" offices.

Manufacturing is moving in the opposite direction, however. Chips and chip-based products, valuable but small, are cheap to ship. Silicon Valley–based semiconductor, computer, and peripheral equipment companies have established production facilities in Oregon, Idaho, and Texas, far from suppliers or vendors (in some cases these may be other divisions of the same firm), because the costs of shipping are minor compared to the remote sites' production cost advantages. In fact, since the earliest days of transistor production, semiconductor producers have shipped partially completed components to Asia for assembly before flying them back to the United States for testing and distribution.

Improvements in the transfer of information, particularly computer data, have made the remote production facilities viable, while these same advances have permitted the centralization of management. Semiconductor producers like Fairchild and Texas Instruments tie together their domestic and Asian plants via private satellite-linked computer networks. Similarly, finance and insurance corporations have moved data entry (typing) pools from their downtown offices to the suburbs, to rural communities, and even, like American Airlines, overseas, linking them to headquarters electronically.

This trend toward deconcentration has not occurred only in the high-tech and financial industries. Nationwide, in the 1970s, the rate of employment growth in nonmetropolitan areas, especially those not adjacent to urban centers, surpassed metropolitan job growth. With "WATS-line" long-distance phone service, computerized management systems, and electronic mail, branch plants virtually anywhere in the United States can stay in close contact with their company headquarters.

■ For twenty-eight years, the Ford Motor Company's Milpitas, California, facility was an economic mainstay of the San Jose area. When it closed in 1983, it had been employing 2,400 workers. Paying union wages for blue-collar work, the plant represented a major economic opportunity for Chicanos and other minorities in the Santa Clara Valley; however, in late 1982, Ford announced that it planned to close, due, it said, to falling sales. And in late 1984, a Silicon Valley developer released his plans to convert the site into a high-tech industrial park.

Even before the plant ceased operation, the United Auto Workers and Ford initiated a comprehensive training and reemployment program for displaced workers. Hundreds of workers have received help through the joint effort, which included both courses and on-the-job training in such high-tech specialities as computer repair, semiconductor mask design, and microwaves.

Political leaders across the political spectrum view high-tech production as a solution to both the existing and the potential unemployment problems which high tech is at least partially responsible for creating. In general, Democrats favor a government role in targeting investments, while Republicans emphasize tax benefits designed to reward private initiative. Both parties favor the attraction of high-tech production to communities where jobs have been lost, and the migration of workers to centers of high-tech development.

A key element of all high-tech programs, regardless of the scheme for economic planning, is retraining. If workers displaced by the shift to high tech are to be reemployed, they must be given new skills. But the UAW–Ford Motor Milpitas program, which is one of the nation's most successful adjustment schemes, demonstrates that retraining is at best only a partial solution.

The Ford workers had key advantages not available to most workers cast off by declining companies and industries. First, they were situated at the edge of Silicon Valley industrial growth, so they didn't have to migrate to find work. Second, their union had both the strength and the concern to continue to represent the interests of its laid-off members. And third, the UAW had built into its Ford contract a system of supplementary unemployment benefits, which has helped many workers bridge the financial gap into new careers.

Yet even the UAW-Ford program had its shortcomings. To ensure the success of its high-tech training programs, as well as to protect unemployed workers from the emotional damage of a second layoff, the program screened its participants. Those without basic skills were offered remedial training in basic math or courses in English as a second language. Many ended up in landscaping or auto service, not high tech. This, of course, is not the fault of the retraining program, but it illustrates the difference between high tech and heavy industry. To reach skilled employment levels in high tech, one needs a greater command of the English language and a higher level of education than is necessary for advancement in traditional blue-collar work.

The other major shortcoming is obvious. Workers leaving secure union-scale positions at Ford moved into nonunion jobs with limited job security and wages roughly half of what they received at Ford. Typically, earnings fell from $12 per hour, plus an impressive benefits package, to $7 per hour. Union officials were optimistic, however, that supplemental unemployment benefits payments would maintain the retrained worker's incomes long enough for them to climb the high-tech pay scale. Until a worker's benefits are exhausted, his earnings are maintained at 95 percent of his Ford wages.

Surprisingly, the UAW found that anti-union high-tech companies had no hesitation about hiring former union members. Apparently, employers considered a record of stable employment to be more important than potential support for unionization at their new jobs.

The UAW, while proud of the Milpitas program, does not agree with those who see high-technology industry as an economic panacea. Its "Blueprint for a Working America" urges programs to create jobs at prevailing wage levels and calls for a balance between strengthening traditional "smokestack" industries and high tech. The UAW's commitment to retraining also points out a conflict between the immediate and future needs of its members. Benefits, such as supplemental unemployment insurance and retraining, cost American automakers dearly. Management-oriented analysts argue that the pay packages which include those benefits represent a severe handicap to the U.S. auto industry in its attempt to compete with Japanese and other foreign auto firms.

■ In the early 1980s, Electronic Arrays—a Silicon Valley semiconductor manufacturer owned since 1978 by Nippon Electric—introduced computerized equipment for bonding threadlike gold wires to semiconductor chips. Not only did the automated assembly line displace a number of workers but it reduced the skill required of those who remained. It takes about two weeks to train workers to operate the automated equipment, as compared to two to three months under the old, manual system.

While many workers are threatened with unemployment caused by the application of high technology, many more face changes in the way they work. It may appear that the brave new workplace is the direct, unavoidable result of new technologies, but owners and managers have shaped those technologies and determined how they are applied.

In some situations, employers simply substitute high-tech machines for one or more workers; in most cases, however, they systematically use new processes to reorganize the division of labor in their businesses. Work that in the past was done by a collection of uniformly skilled workers is dissected into new sets of distinct tasks. Employers assign jobs requiring skill to only a handful of well-paid workers, while giving everyone else as little responsibility and money as possible.

This "scientific management" approach—minimizing production costs by redividing the labor process—is not new. Frederick Taylor popularized it nearly a century ago. Although many labor organizers today associate "Taylorism" with time-and-motion studies of individual workers, scientific management has historically focused on the shape of the entire workplace.

Today, high technology gives managers the tools with which to restructure labor as never before. For example, sophisticated switching technology permits telephone companies to reorganize customer service. When you call in for certain kinds of help, you at first get someone trained and paid to handle routine inquiries or complaints. Then, if she (it is usually a woman) cannot help, you are routed up-circuit to a better-paid employee or supervisor.

The same telephone companies have traditionally employed large numbers of skilled technicians to keep their networks functioning properly. Admittedly, the recent introduction of microprocessor-controlled electronic switching systems opened up some positions for programmers at the upper end of the occupational hierarchy; yet the automated, self-troubleshooting switching equipment has rendered useless the skills of many technicians. Experienced troubleshooters were given an unpleasant choice: They could either quit, or accept lower-paying technical positions requiring less skill.

Throughout the economy, skilled middle-income work is gradually disappearing. Although some skilled workers, as well as new entrants into the labor force, will be able to find highly paid professional positions, most will follow the telephone technicians, machinists, and displaced auto workers into less remunerative work. Obviously this is a problem for the affected workers, but its implications go much deeper.

If present trends continue, America's "dream technology" of high-tech computers, chips, and communications may destroy the social dream of a middle-class society. The new working poor—semi-skilled office, service, and production workers—are joining the gradually growing ranks of the unemployed at the bottom of the economic ladder.

Throughout this century, unions have played an important role in maintaining middle-class incomes, both by bargaining for decent wages for their members and by threatening to organize nonunion companies and agencies which did not pay competitive wages. Today, the labor movement remains the greatest defender of middle-class pay levels, but unions are having difficulty hanging on to their wage scales and benefits. Not only must organized labor compete in the labor market against nonunion labor, but workers must also match the performance of automated equipment.

■ In addition to using computerized systems to cut costs by altering skill levels, employers are applying computer technology to increase the rate at which workers function. It is ironic that the video displays on computers are known as monitors, implying that the fundamental person-machine interaction is one of the person monitoring the machine. This may be true for many professionals and managers, but the average worker in a computerized workplace is much more likely to be monitored *by* a machine.

It is technically simple for a computer to record, analyze, and transmit immediately to supervisors data on the pace of employees—whether they be typists, sales clerks, or assembly-line operatives. And just as easily, performance can be analyzed at the end of the day, week, month, or year. At United Airlines and TWA, for instance, management uses the companies' in-house automated reservation systems to count the number of reservations sold by each ticket agent. Similarly, telephone companies and telephone marketing firms monitor the "average work time" of their operators.

The technology is new, but the concept is an old one, known as "speed-up." Many employers believe they can improve productivity by carefully monitoring the speed of their workers, rewarding the fast ones and penalizing or firing the slow ones. Anyone who has ever worked under such pressure knows that fancy electronic devices, just like pushy supervisors, tend to produce stress. Faster rates of production, in the absence of labor-saving techniques and equipment, merely increase the level of exploitation.

In the long run, speed-up can only marginally increase productivity. Those managers who rely on this technique are taking a chance, for it may actually cut productivity by disrupting work patterns. Conscious rebellion, in the form of strikes or sabotage, is possible; or sped-up employees may merely cut corners—potentially damaging not only the product but their machines or themselves—or take time off.

Quality, which often is not measured by the machines, frequently suffers as quantity increases. For instance, telephone installers who are dispatched and monitored by computers tend to hang lines along the shortest path, abandoning their old, slower techniques of skillfully concealing service drops.

In some instances, computerized systems are used to pace employees, rather than monitor them, in order to maintain or increase production rates. In other words, in many automated factories, workers have no control over the flow of work. The growth of machine-paced production has actually reduced the need for the older methods of speed-up: exhortation, financial promises, and threats.

Centralized communications systems are designed to set the pace of operators, whether they work for phone companies, airlines, or catalogue companies. Sitting in centralized "boiler rooms," they are fed a call from virtually anywhere as soon as they complete the last one. As Linda Rolufs, an operator at Canada's BC (British Columbia) Telephone, which is owned by U.S.-based GTE and run in the same fashion as American phone companies, reports: "The computer connects the operator to a new call as soon as her other customer is off the line. . . . If your desire is to process as many calls as possible in the shortest possible time, then I guess you could say that this is an increase in productivity. . . . What used to be a human work environment has become like an assembly line."

■ In 1983, the National Association of Working Women (also known as 9 to 5) operated a hotline for video display terminal (VDT) users. During its six months of operation, the hotline was swamped with more than six thousand complaints. Most of the callers reported eye discomfort, musculoskeletal pain, and stress; but 9 to 5 also identified fifteen possible clusters of problem pregnancies in computerized offices. The Association reported that during 1979–84, half of the pregnancies known to have occurred among the three hundred video display workers at United Airlines' San Francisco offices resulted in adverse outcomes, including miscarriages, birth defects, and stillbirth.

Fears that video units may emit dangerous levels of radio-frequency waves or even more dangerous X-radiation have prompted many pregnant VDT workers to buy and wear lead aprons, yet researchers who have studied emissions from the typical display unit manufactured today argue that the emissions are no stronger than ambient (background) radiation levels. Standards set by testing firms such as Underwriters Laboratory require that units be designed so that they will not give off X-radiation even if components malfunction. There may be some older, more dangerous machines still in operation, and there may be hazards not yet understood by scientists, but the problems of VDT use probably have nothing to do with radiation.

Most video display terminals have been introduced into workplaces with little thought about how people would interact with them physically. Now, however, specialists in ergonomics, the relatively new science of adapting machines to people, are reporting that most automated workplaces suffer from improper lighting, nonsupportive seating, and keyboards and screens at the wrong height or angle. By redesigning equipment, furniture, and lighting, experts can reduce or even eliminate many of the complaints of VDT operators.

One might think that employers are employing ergonomists only to improve conditions for the growing number of professionals who work in front of video terminals; however, some companies have been willing to invest substantial sums in new equipment for employees at the bottom of the hierarchy. For instance, American Airlines bought special chairs, which retail at an estimated $700 each in the United States, for its Barbados data entry clerks, who earn only $1.75 to $3.00 per hour. In this case, at least, management believes that more comfortable workers are likely to be more efficient.

But workers in even the best designed offices can suffer from fatigue, stress, and such stress-related ailments as heart disease if they labor all day long at video units. In effect, concentrated VDT work is a form of speed-up. As *Fortune* writes: "Not quite keeping up with your computer can in itself be frustrating. But if you aren't sure you're keeping up with your fellow workers either, and your boss knows it, that's a more serious strain. All the boss has to do is push a key or two at his own VDT, and the statistics of performance are clear. Though ergonomists identify this kind of stress, only management can lessen it." This is the same stress described by Linda Rolufs, the Canadian phone worker.

Workers aren't waiting for management to solve their problems, however. Unions are beginning to insist on frequent breaks and limits on VDT use in their contracts. Rolufs's union, the Telecommunications Workers Union of Canada, fought grievances on behalf of five pregnant operators who requested non–video display work. Several states are now considering workplace legislation which would establish standards for breaks, length of work day, and ergonomics. In Europe, such laws are already on the books.

Thus, the evidence so far indicates that the growing number of complaints about video units result from the new workplace organization that computerization has made possible, rather than from computer technology itself. It is possible, over the next five years, that flat panel displays, which are similar to the liquid crystal displays now found on many watches and pocket calculators but larger, will take over the market for new video terminals. These new displays are reflective, low-power devices that will emit much less radiation than the already low levels associated with VDT's, but their screens are dimmer. As long as low-level workers are kept slaves to their terminals, health problems will continue.

■ Raw speed-up is merely the simplest feature of a more general trend, the use of high technology to give "scientific" managers increased control over the workplace. Word processors, numerically controlled machine tools, "point-of-sale" computers, financial data terminals, and other computerized work stations have all been designed to provide management with timely, detailed data on the nature of the work process, not just the pace of individual workers. In general, as management learns more, workers lose their autonomy. When top managers centralize their data on the flow of work, they can further reorganize the workplace to increase productivity or to lower costs, even over the objections of workers whose skills were once highly valued.

At present, management is often unwilling to share power over the introduction of technology with its workers. But workers must demand such power. Unless they can fight successfully to control technology, employers will use the technology to control them, undermining their working conditions.

In Europe, and particularly in Scandinavia, labor unions have bargained over the introduction of high tech for several years. In America the International Association of Machinists has proposed a ten-point Technology Bill of Rights, which they want appended to the National Labor Relations Act and other federal labor legislation. More of an organizational program than a realistic piece of legislation, the Machinists' Bill of Rights would mandate that "workers, through their trade unions and bargaining units, shall have an absolute right to participate in all phases of management deliberations and decisions that lead or could lead to the introduction of new technology or the changing of the workplace system design, work processes and procedures for doing work, including the shutdown of work, capital, plant and equipment" (Point 7). Point 8 would promise workers the right to monitor control rooms while preventing management from using computerized systems to monitor individual workers.

Of necessity, much of the bargaining done by unions so far has been to protect or retrain workers in the face of automation. Immediate working conditions, such as the length and frequency of breaks from video work, have also drawn attention. Overall, however, workers and their representatives have had little to say about the introduction of new technologies.

Nevertheless, in the long run the new technologies can be used to create a new, more positive workplace, in which individual workers have more autonomy while employees as a group have a much greater influence over the way production is handled or services are rendered. Computers need not lead to hierarchical specialization.

Training workers to do many jobs is a simple initial step. The Canadian Telecommunications Workers have insisted that operators learn more than one task, and even Apple, at its automated Macintosh plant, trains employees for several positions. The ability to work at many stages of a process not only makes work less routine; it also gives workers a greater understanding of the overall work process. Employers can exploit that knowledge to improve productivity. When workers recognize the value of their knowledge, however, they can work toward increased power over their entire work situation.

Earlier forms of factory automation trapped workers —as Charlie Chaplin showed so brilliantly in *Modern Times*—at one task per work station. Today's flexible systems, such as programmable machine tools, could permit a return to the pride of craftmanship. Instead of operating a tool repetitively, a factory worker could use the same tool to carry out several processes rapidly on one product. Workers, if plugged into the process of programming the tools, could ensure quality output, so justifying a higher level of both pay and status.

In one of our offices, everyone does his or her own typing. This means that no writer spends all day at a terminal, and we each carry out a variety of functions, such as interviewing, reading, and filing, during a typical work day. Not all offices can divide work this way, but most that could are still organized according to the dictates of "scientific management."

The people who handle telephone complaints at credit card companies or the phone company could be given the training and authority to handle unusual problems, rather than referring these "upstairs." Today's computerized billing systems make it easy for any authorized employee to call up a customer's file. At slight expense, both the employees' morale and the customers' satisfaction would rise appreciably.

Innovative, egalitarian approaches to the introduction of new technology are by no means a panacea for all the workplace problems of industrial capitalism. Neither the encouragement of input into process design nor cross-training can be substitutes for overall industrial democracy. But such approaches at the very least give workers whose knowledge and autonomy have been downgraded for decades a sense of their abilities and a taste of increased power over their own lives.

Low-cost microelectronics equipment and technology have made it possible for some people to own their tools, materials, and technology, and so avoid the exploitative relationships of large corporations and agencies. For instance, today virtually anyone who wants to write and market software can purchase his or her own microcomputer. Even high school students are now running their own programming businesses. Before the development of microcomputers, most programmers had to work for large organizations; they had little autonomy, and in many firms programming resembled assembly-line work. But it is important to recognize that such opportunities are limited to a small group of elite professionals. Then, too, since marketing channels are easily monopolized, many who successfully develop new software are not in a position to reap the full benefits.

In the long run, if one can believe the projections of companies specializing in the booming new field of "artificial intelligence," virtually any professional occupation can be automated. So far, data processing, in the conventional sense, has been used to automate only the routine functions of the human mind. Today, researchers are attempting to develop machines which can *reason* as well. They are designing and testing both hardware and software to formulate, recognize, and comprehend human language; to emulate eye-hand coordination; and to diagnose problems and put forward solutions.

Diagnosis and problem-solving routines, known as "expert systems," are elaborate computer programs designed to absorb the accumulated knowledge of experts in specific fields, not only to solve known problems but to learn how to solve new problems never faced by the human experts. At present these are used to supplement the expertise of doctors, lawyers, geologists, and so on, but they may eventually displace human experts, at least in applications where the results are somewhat predictable. Computers are approaching the capacity to render such expert systems economically viable if researchers can succeed in making them work more effectively and reliably than their human counterparts.

Some human experts are understandably concerned. Chip designers at one major Massachusetts minicomputer company actually refused to share their knowledge with researchers putting together an expert system to do chip layout, just as craftsmen refused to share their skills with scientific management's pioneers decades ago.

At a time when many Americans are espousing a new reverence for human life, our regard for human labor is falling. For the average worker today, work is no longer a fulfilling enterprise, and for an increasing number of people, labor is an inadequate source of income. Our economy may flourish as technology downgrades human activity into a mere factor of production, like silicon, electricity, or money; but our society will certainly suffer.

Essentially we face the same dilemma that confronted our predecessors during the first industrial revolution. Can we create a society which rewards creative thought and industrial innovation while at the same time providing economic and social opportunities for all?

Chapter Six OUTSIDE THE ELECTRONIC COTTAGE

■ ■ ■ ■ ■

To the self-appointed heralds of the information age, success today is no longer symbolized by a "chicken in every pot" or "two cars in every garage," but by a "computer in every cottage." Millions of Americans now have personal computers. As the prices of microcomputer hardware and software continue to fall, home computers, they suggest, may become as common as televisions.

For a sizable minority of the population, the electronic cottage promises unprecedented access to information-based services. However, the so-called information revolution is likely to bypass the majority. In fact, since one's position in society is increasingly defined by access to ideas, knowledge, and data, a vast new stratum of information-poor is emerging.

The personal computer, when connected to the nation's telephone network, is a key that can open the distant doors of retailers, banks, schools, libraries, and correspondents. With the proper equipment, anyone can quickly send electronic messages back and forth across the country from the comfort of his living room, study, or office. High-tech information services, known as videotex, are in their infancy, but major corporations are betting vast sums of money on the enticing convenience of the electronic cottage.

Though most Americans will soon, if they don't already, interact with computers in their daily lives, it is unlikely that even a majority of U.S. households, in the foreseeable future, will have their own machines, complete with the hardware and software for two-way communications. Today, only about 3 million homes are equipped to send and receive computer data.

Just as many poor and elderly people were left stranded when the automobile pushed the trolley off the streets of urban America, large numbers of Americans will discover that their traditional channels of information—the mail, the voice telephone, branch banking, and public libraries—are becoming more expensive, less reliable, and harder to find and use. Only one-way sources, such as television and radio, will remain universal.

Giant communications corporations are also using new technologies to centralize the flow of news and ideas. Network news departments and on-line data bases, for example, owe their rising influence to breakthroughs in communications and data processing; here the small and local media simply cannot compete.

High tech, therefore, threatens our democratic traditions. The media monopolies are by no means monolithic, yet the concentration of information supply does limit the influence of groups with new or unorthodox viewpoints. It is true that low-cost, microelectronics-based communications and computer technologies make it possible for members of a technological middle class to own and operate their own, independent information ventures. But with limited resources, such channels will reach only a small portion of the population.

■ For most people, and organizations, the telephone system is their link to the world of automated information services. Although the costs of long-distance communications are falling, local phone rates are rising rapidly in most states. Fewer people will be able to afford telephones; those with a subsidized "lifeline" phone service will be limited to a few calls each month.

Since the late 1960s, the rise of new microelectronics-based communications technologies has prompted a series of federal policy decisions designed to encourage competition in the telecommunications industry. Beginning with FCC rulings allowing non-AT&T equipment and transmission lines to be plugged into the Bell System, the process culminated with the January 1984 dismantling of the American Telephone and Telegraph Company itself.

The new technologies are altering both the cost structure of supplying telecommunications services and the make-up of consumer demand. With the establishment of microwave networks and satellite relays in recent years, the cost of providing long-distance hook-ups has declined, particularly in relation to the installation, maintenance, and operation of local phone systems. Meanwhile, the growth of computer-to-computer communications has led to the concentration of volume in the accounts of a few customers.

Large corporations, which need high-volume, long-distance communications, have sought to take advantage of the reduced cost of providing long-distance links. A few have set up their own networks, bypassing entirely the common carrier telephone system; but most pressured the federal government to open up long-distance service to competition. The first crack in the monolith occurred in 1969, when the Federal Communications Commission (FCC) gave preliminary approval for MCI, Inc., to establish an alternate long-distance microwave channel between Chicago and St. Louis. Within a decade, more than a dozen major firms were offering long-distance voice and data transmission services.

Despite their cheaper rates, the alternate systems have not been able to capture a large percentage of AT&T's business. As late as 1983 they held only about 8 percent of the long-distance market, largely because subscribers had to dial a long series of numbers to take advantage of their lower rates.

At the urging of large corporate users of telecommunications, the Justice Department pursued its anti-trust case against AT&T to encourage further competition in long-distance service. In January 1982, the Justice Department announced a settlement with AT&T, in which the nation's phone monopoly agreed to divest its local operating companies into regional corporate empires.

As the settlement is gradually being implemented, AT&T is losing its favored position as the standard long-distance service. With "equal access," customers can use another long-distance carrier without having to dial extra codes. The big customers will not all abandon

AT&T, however. The alternate phone networks are being forced to raise their rates as they receive equal access, while competition is bringing down AT&T's long-distance rates.

Although the break-up of the Bell System is advantageous for large corporations, it is driving up the phone bills for the average residential telephone customer. When AT&T ran the entire, regulated system, it used income from its long-distance department to "subsidize" the operation of its local networks. Now, the regional phone companies are seeking, and usually receiving, massive local rate hikes from state regulatory agencies across the nation to make up for the lost long-distance revenue.

Consumer and labor organizations dispute the fairness of the accounting assumptions made by the telephone companies and the FCC. They argue that local ratepayers, by providing the local systems that serve as outlets for long-distance calls, are actually subsidizing large, corporate customers. In other words, if long-distance companies did not have access to local phone systems, they would have to build their own, at enormous cost.

However, financial practicality, not fairness, is shaping the future of American telecommunications. The same technologies which are enhancing the efficiency of American telecommunications are forcing regulators to make local residential and small business customers pick up a portion of the tab previously carried by the corporate sector. With the latest switching equipment and microwave transmitters, large companies and institutions can bypass local phone networks, directly connecting to long-distance carriers. Many have threatened to do exactly that if legislation or an FCC ruling requires them to pay more to local companies. When firms completely bypass local networks, they don't have to pay any local charges.

Not only are average customers paying more for their "plain old telephone service," but across the country they are bearing the huge cost of rebuilding the entire phone network. The phone system is rapidly being transformed from a voice-only network into an "integrated services digital network" (ISDN). To provide high-speed computer communications, local phone companies throughout America are discarding working equipment that transmits perfectly satisfactory voice communications in order to install instead the latest in high-tech switching and transmission equipment. Just as vastly overpriced nuclear power has been foisted on consumers by gas and electric companies, phone companies are choosing to let consumers pick up the tab for the information age. Nobody has asked whether the consumers want to finance the new information services.

Thus, while the nation's telecommunications network is being reorganized and rebuilt to meet the needs of a corporate-designed information society, skyrocketing local phone rates will exclude large numbers of people from full participation in that society. The nation's largest phone workers' union, the Communications Workers of America, reported in 1983: "According to AT&T's own data, a doubling of the basic telephone rate would reduce the number of households with telephone service from 92 percent of all households to 84 percent; a tripling of rates would reduce the percentage of households with telephone service to 71 percent, dropping the number of households with service from the current 74 million to 57 million." Not surprisingly, the poor, the elderly, and the rural will be hardest hit.

To maintain the American tradition of a universal

phone system, Congress, the FCC, and their state counterparts need to act immediately to provide subsidies or revise their regulatory strategies. Minimal phone service can be retained through the expansion of the "lifeline" service; that still leaves the nation's poor far from the computerized cottage.

■ Today, anyone with a communicating computer can transmit electronic messages over the phone directly to the computerized "mail slot" of an individual or even to an entire distribution list. With electronic mail, letter writers need not even take the time to print out and stuff computer-composed letters into envelopes, and they can send the same or similar messages to numerous correspondents. Since electronic mail is transmitted instantaneously, it beats not only the U.S. Postal Service (USPS) but also the rapidly expanding package express services as well.

While voice communications are often preferable to electronic messages, the ability to deliver an electronic memo frequently eliminates the inconvenience of telephone tag. Two people with conflicting schedules can carry out a complex, rapid two-way interchange without ever talking directly to each other.

Some of the electronic mail services, such as the U.S. Postal Service's ECOM (Electronic Computer Originated Mail) or MCI's MCIMAIL, convert computer messages into hardcopy for delivery through the regular mail or by courier. These services may prove important to bulk mailers, although they otherwise have limited value. Fully electronic mail, however, should catch on by the end of the decade. By then, enough people and organizations will have electronic mail slots at MCI, AT&T, the Source, Compuserve, or other E-mail providers to enable businesses and affluent computer users to conduct normal correspondence electronically.

Electronic mail can only serve that portion of the public that has communicating computers and phones or other network connections. In addition to normal phone charges, one must also pay the electronic mail service each time a message is dispatched. When the information-rich switch to electronic mail, the U.S. Postal Service and those Americans who rely upon it will suffer. As people desert the universal mail system, economies of scale—that is, efficiencies in the use of personnel or machines due to high volume—will disappear. Without additional subsidies, the mail service will deteriorate and rates will rise more rapidly. Since corporations and those of above average means will be mailing electronically, they will no longer lobby on behalf of the Post Office, so that Congress is unlikely to help out with budgetary assistance.

Conceivably, the postal service might be allowed to set up its own fully electronic service, allocating its revenues to maintain a mixed system of both electronic and hardcopy delivery. But this contradicts present policy in three ways. First, the USPS is currently not allowed to engage in electronic communications; second, the federal government is encouraging competition in electronic mail services; and third, the postal service is required, over a period of time, to make each of its services self-sufficient.

Giving the postal service a universal monopoly on inter-customer (meaning outside a single institution or corporation) electronic mail might benefit the average American, but it would probably reduce services for those people and companies already utilizing privately owned electronic mail systems. And those are the users who, through their above average concern and resources, have the most influence over telecommunications policy.

■ While America has a strong tradition of universal postal and telephone communications, financial services are considered a privilege. The new information technologies are widening the gap between rich and poor bank customers, and it is unlikely that proposals to equalize services will be taken seriously.

In modern America, number-shuffling, not cash-handling, is at the heart of banking. Financial institutions automated their internal information systems years ago, and today they are automating their dealings with customers. Electronic cash-vending machines are merely the tip of the iceberg. More important, banks and financial service companies throughout the country are testing and initiating programs through which they will allow the owners of data terminals and personal computers to check their accounts, transfer funds, and pay bills from home. Frost & Sullivan, the market research firm, projects that by 1991 one in ten households will bank electronically from home.

For several years, some banks have promoted telephone bill-payment (TBP) systems, by which a customer with a touchtone phone can dial in the information for bill payments to cooperating vendors, such as utilities. Since TBP provides no instant feedback or accounting mechanism, it is already being eclipsed by computerized systems, which require a video display terminal.

In December 1983, the Bank of America, which pioneered branch banking decades ago by placing an office on virtually every main street in California, launched one of the country's most ambitious computerized home banking programs. With a communicating computer or terminal, one can dial into the Bank of America's system, known as HomeBanking. Using the proper identification number and instructions, a customer can view his or her account balances, transfer funds between accounts, and pay bills to other financial institutions, credit card companies, and utilities, as well as to participating hospitals, fitness clubs, and department stores. In its first year, the bank signed up fifteen thousand customers.

For consumers, home banking is an obvious convenience. One can carry out most common banking tasks, except nonautomatic deposits, from the home or office at any hour. Transactions are posted quickly. Customers can easily monitor their savings, checking, and credit account balances between printed statements (U.S. law requires that they be sent printed records of all transactions). In the HomeBanking system, payments can even be scheduled ahead of time.

Banks expect large numbers of customers to bank from home once the services are well established and the costs fall. Many personal computer users will begin home banking when they subscribe to multi-function videotex.

Home banking benefits financial institutions even more than it helps their customers. Most banks and savings institutions already have computerized accounting procedures, so for relatively little additional investment, they can expand those systems to serve customers at home. Home banking effectively automates "front offices," reducing the required number of human tellers and, eventually, the number of brick-and-mortar branches. The savings in labor costs and overhead are potentially enormous.

Furthermore, banks plan to use home banking to compete for affluent customers, who are a key source of income. Banks earn about 90 percent of their retail profits on the business of the wealthiest 20 percent of their customers.

Meanwhile, those customers without a communicating computer will find that their banking fees are rising faster than those who bank from home. This is consistent with another trend in banking: banks offer high interest rates and free services to clients with large balances, while the remaining customers must pay escalating service charges and earn little or no interest on their funds.

Furthermore, the growth of home banking, coupled with the installation of automated tellers, will make it difficult for the average customer to do business with a human teller. In 1983, when the Bank of America was preparing its home banking system, it started converting some of its branches into teller-less offices. In those locations, automated teller machines have replaced tellers and other personnel have been shifted to centralized main offices.

To top this off, one week after it initiated HomeBanking, the Bank of America announced plans to close 120 of its nearly 1,100 branches. Several years ago the Bank of America ran a TV ad campaign stressing the convenience of its many branches. Today it appears to be concerned primarily about convenience for upscale customers who can afford to bank from home.

The Bank of America is singled out here not because it is less moral than its competitors, but because it has always been a pioneer in retail banking. Other financial institutions are initiating similar programs. Possibly some bank or savings institution will find its market niche by serving those customers who have been deprived by the front-office automation of retail finance. Or more customers may join cooperative credit unions. Overall, however, it looks as though little can be done within the present structure of the American banking industry to reverse the rise of this two-tier financial system.

■ Most people probably expect banks to cater to the rich, yet few recognize that the growth of automated information retrieval services is likely to create a two-tiered information market, in which the information-rich have privileged access to news and research data.

Microelectronics technology has made possible the development of on-line data bases—computerized information services which combine the timeliness of newspapers with the depth and breadth of research libraries. With a terminal or communicating computer, a paying subscriber can call into a number of general service data bases, such as the Dow Jones News/Retrieval, the Source, Dialog, and Nexis. In their electronic memories, these services store stock quotes, bibliographies, newsletter abstracts, patent filings, and even the texts of entire newspapers. Television and radio deliver information to us just as rapidly, but on-line data bases are "interactive." From his home or office, a subscriber can select precisely the subject matter he wishes to receive.

Like electronic mail and home banking, on-line data bases are directly available to the minority of the population with communicating computers. In fact, they are effectively open to very few Americans, since electronic file searches are expensive. It costs $50 to join Dow Jones News/Retrieval, one of the biggest services, and as much as $1.20 per minute for connect time. Not surprisingly, most Dow Jones subscribers are corporations or business people.

Public libraries help the average reader by subscribing to data bases and providing indirect access; in addition, libraries throughout the country are joining computerized cataloguing networks. But both approaches are becoming increasingly costly. For example, the On-line Computer Library Center, which provides inter-library and cataloguing services, told the public library in Rapid City, South Dakota, that it could expect a 68 percent increase in its telecommunications bills if proposed rate increases were approved. In fact, the rapid rise in telephone prices we have just analyzed may force some libraries to drop their computerized services or make substantial cutbacks elsewhere.

Phone costs aren't the only strain on library budgets. Our tradition of free public libraries must wait behind police, fire, and transportation when local governments re-divvy their inadequate budgets each year. Libraries are not merely having difficulty providing access to the new high-tech information services; they are being forced to cut back their traditional offerings. Staff, hours, and services have all been curtailed in many cities. In New York City, for instance, the branch library system employed 1,300 people in 1970. By 1981, despite growing needs, the total was 1,000, and less than a third of the branches were open on Saturdays.

As corporations and affluent individuals take advantage of computerized information services, America's great public library system is likely to lose even more political and financial support. Unless a conscious effort is made to balance the scales between the information-rich and the information-poor, average Americans will wake up one morning and find that the information revolution has passed them by.

■ Although on-line data bases are used by a relatively small number of researchers, their impact on society as a whole is potentially great, as reporters, commentators, and others in the mass media grow dependent upon their timeliness and convenience. However, since it is costly to establish an electronic library large enough to be useful, system operators have designed their data bases to serve those customers best able to pay—large corporations.

Nexis, for instance, is a remarkable source of information. Its computers store the full texts of articles from at least seventy-five publications. A subscriber can locate reports on almost any subject merely by keying in a few words related to that subject. He can then print out or dump into his computer a portion or the entirety of any of those reports.

The Nexis data base includes the nation's leading daily newspapers, news magazines, and the wire services, but most of its sources are business and trade magazines. It ignores, of course, thousands of small newspapers, newsletters, and magazines, and does not even index popular ideologically oriented magazines such as *Mother Jones, The Nation, The National Review,* or *Reason,* all publications which cover subjects and offer views not frequently found in most business and news periodicals.

Those who operate centralized on-line data bases do have the opportunity either subtly or directly to bias their systems. The history of computerized airline reservation systems—specialized data bases serving travel agents—illustrates this danger.

More than 80 percent of America's travel agencies are tied into the country's two largest reservation networks, Apollo and Sabre. United Airlines operates Apollo; Sabre is part of American Airlines. Both systems provide flight listings that favor the runs of their parent companies. Since favorable displays mean more bookings, smaller airlines have gone to the Civil Aeronautics

Board, the courts, and Congress to win equal treatm

Those dissatisfied with the existing data base sy tems may eventually establish their own alternativ systems. The costs of personal computer equipment and even mass storage systems are falling so dramatically that the construction of small, nonprofit systems will soon be feasible. However, unless such systems can find volunteers to help convert large volumes of hard-copy text into computerized form or include only those materials produced on word-processing equipment, the operating costs will be prohibitive. Income is likely to remain low, since nonbusiness library users are generally not inclined to pay for research aids. At best, alternative data bases will become the late 1980s counterpart of the "underground" press of the late 1960s. Though they may reach only a small number of people, they could force the big systems to include more points of view.

■ Communications satellites, one of the silicon era's most far-reaching technologies, also tend to centralize control over information as they bridge entire continents. Satellites are at once the products and the tools of the microelectronics era. Without computers and solid state components, the civilian space program would have been impossible. In fact, NASA joined the Pentagon as a key early customer, providing a market for integrated circuits before they became commercially viable.

The USSR began the space race with Sputnik, but the United States, easily the leader in microelectronics, has dominated the commercial use of space. Those countries with satellites and ground stations launched by NASA and built by firms such as Hughes Aircraft and Ford Aerospace are dependent upon the continued goodwill of the United States. Joerg Becker, a West German telecommunications policy expert, reports that the Pentagon can switch off Indonesia's national satellite system, Palapa, with "one push of a button." He charges that the United States considered cutting off Iran's link to Intelsat, the international satellite consortium led by the United States, during the 1979–81 hostage crisis. "This would have shut down Iran's telephones, its television system, electronic funds transfers, and flight reservations."

The relaying of television signals around the globe is probably the most dramatic and familiar use of commercial communications spacecraft. Only a few years ago, it was impossible to view the Olympics direct from Sarajevo or watch ABC anchorman Ted Koppel grill, live via satellite, such personalities as Mohammar Khadafy, Ferdinand Marcos, Desmond Tutu, or Soviet commentator Vladimir Pozner.

The high-cost transmission of television news via satellite does not prevent those who operate small radio stations or publish local newspapers from covering the same subjects. But how many Americans will listen to two professors debating the niceties of international law or read an essay on the communal history of Lebanon when they can watch, almost live, the siege of Beirut? In America, news is a form of entertainment. Those who disseminate the news without the polished presentation and exciting film footage of high-tech broadcasting enjoy freedom of the press, but they play to small audiences.

However, as new techniques and chip-based communications equipment are introduced, the cost of satellite communications is falling while the number of channels is increasing. Potentially, technology could open up new sources of information, even if it has so far meant only "more of the same." The Cable News Network, for example, provides newsaholics with access to TV journalism at any hour, but its offerings mimic those of the major broadcast networks. Meanwhile, cable television's public access and ethnic channels suffer from inadequate funding.

- Political activists have long argued that "Freedom of the press is guaranteed to those who own one." In the silicon age, there is a high-tech counterpart to the independent press. Anyone with a communicating computer and a telephone can set up a Bulletin Board System (BBS), which combines the various features of electronic mail, on-line data bases, and neighborhood bulletin boards. Most bulletin boards are free to anyone with a computer terminal. They are limited, however, by the number of phone lines—usually one—hooked up to the host microcomputer.

Simple bulletin board systems have proliferated like wildfire throughout the United States as computer users found that they offer a pleasant, practical, and inexpensive way to stay in touch even when separated by time and geographic distance. Special-interest bulletin boards on topics as diverse as astronomy, amateur radio, flying, and fantasy role-playing have sprung up. Computer Professionals for Social Responsibility operates one as a forum on arms control; and Wellnet, which uses equipment donated by Apple Computer, gives disabled people a medium for finding or sharing appropriate goods and services.

These systems have also created headaches for law enforcement officials and raised thorny questions about traditional civil liberties, because some computer users have found they can be used clandestinely to circulate information about how to overcome computer security systems.

As personal computer technology advances, more elaborate networks are developing. One grass roots network called Fido allows its system automatically to pass messages to other systems anywhere in the country late at night when phone rates are low. It is possible with a local phone call to leave a note on a system in San Francisco and have it delivered to a system in New York by the next morning, for little more than the cost of sending a first-class letter.

When bulletin board systems are backed with financial resources, their potential is enormous. For example, for a token launch fee, the Space Shuttle is scheduled to place a tiny satellite called the Packet Radio Satellite (PACSAT) into a low polar orbit in 1986. PACSAT will contain a system open to radio amateurs anywhere on the surface of the globe. With a microcomputer and as little as $700 in communications equipment, they will be able to establish electronic contact within any twelve-hour period. Perhaps the symbol of the grass roots PACSAT approach should be the inexpensive antenna that is needed: Because of the high transmission frequency, a simple twisted coat hanger is sufficient!

Hank Magnuski, a pioneer in low-cost digital communications, is delighted: "Now for the first time you have the potential for global person-to-person communication not based on commercial offerings. It's a tremendous equalizer. A guy who lives in the mountains in the most remote area of the country has, in principle, as much access to information as someone like myself living in the heart of Silicon Valley."

But PACSAT's use will still be limited to technologically sophisticated radio amateurs who are also computer hobbyists. Even with subsidies or continued reductions in communications costs, bulletin board systems will serve only a small portion of the population.

■ In the San Francisco Bay Area, one group of computer-wise activists hopes to bring electronic information services to the entire community. Community Memory, a nonprofit collective of programmers, engineers, and others, has installed computer terminals at public locations in Berkeley such as supermarkets and cultural centers. Anyone can "post" jobs-wanted notices, for-sale ads, political remarks, and other messages on the system. Until Community Memory finds the money to install coin-op devices, participation will be free.

Community Memory's terminals resemble the experimental electronic shopping kiosks now being tested across the Bay in San Francisco, but they differ in their function. Though the kiosks allow shoppers to scan for information and to place orders, they do not allow users to advertise. Community Memory's machines, on the other hand, are designed to carry notices from everyone. In fact, the organization makes a point of exercising no control over the data allowed into the system, since its aim is to build up the sense of community by increasing communications.

Projects like Community Memory are very laudable. But unless affluent donors, foundations, or government agencies make a commitment to public support for such systems, they will have little impact on the increasing trend toward making electronic information a private commodity. After eleven years of hard work by volunteers, Community Memory installed its first three terminals in the summer of 1984. The project may flourish in Berkeley, where the spirit of social innovation matches the momentum of technological change. In the rest of the country, such interactive information services seem to be a long way off for most people.

■ It is dangerous to entrust a new generation of information services to companies that view videotex merely as a sales tool or as a hot new source of revenue. The traditions of American democracy demand that high-tech systems be developed that will not only serve the entire public but also recognize the potential of every individual as a source of information. Today's "alternatives" must be brought into the mainstream.

Yet there is a danger that critics of the existing high-tech media may demand high-tech alternatives in situations where electronic systems do not provide the best service. Public-interest groups are besieged by altruistic high-tech experts wanting to put state-of-the-art computer technology to beneficial use. But, as the Community Memory collective warns, microelectronics technology is merely one of many useful tools:

> The purpose of Community Memory is to aid in decentralizing control of communications and power relations in general. A ditto machine can be used for the same goals, as can the higher technology of the Xerox machine. Community Memory is an electronic filing cabinet, but often a "mechanical" filing cabinet is more appropriate. Computer technology is very sexy. It sells well, and it can frequently be used for tasks which could not be reasonably accomplished another way, but it is expensive, complex, and hard to make reliable. If something can be done adequately without computers, it should be done without computers.

Chapter Seven
THE SILICON RUSH

■ ■ ■ ■ ■

In 1849, treasure-seekers from all over the world flocked to Northern California to mine gold or sell goods and services to the "Forty-Niners." Today, programmers, managers, workers, and others from China to Cincinnati again are making their way toward the Golden Gate. Only this time, rather than seeking the wealth of the mother lode, they hope to strike it rich in "Silicon Valley," where chips of silicon can be worth many times their weight in gold.

The belt of industrial communities at the southern edge of the San Francisco Bay universally symbolizes the promise of the microelectronics era. It was first called Silicon Valley in the early 1970s, when manufacturers of silicon chips became the Santa Clara Valley's major employers. The Valley is home to the greatest concentration of high-tech professionals and enterprises in the world. It is a land where the information-rich, particularly those trained in science and technology, can make both their mark and their millions.

Though Silicon Valley is in many ways unique, planners, officials, and commercial interests throughout the country see the area as a model for industrial growth in the information age. While few other areas can hope to rival the Valley, many have already attracted their share of high-tech facilities. As high tech grows, they will learn the harsh truth behind the legends of Silicon Valley.

Many of the Valley's problems are directly caused by high tech. Others are found elsewhere, but they are significant merely because the residents of would-be Silicon Valleys have been told that the electronics industry has no serious problems. If they study the lessons of the Valley, they can avoid many of the pitfalls of high-tech growth.

■ Maria, a twenty-six-year-old political refugee from Argentina, who chooses not to be known by her real name, found work in Silicon Valley, but she did not strike gold. Maria quit her $4.10 an hour production job at Memorex to have her first baby. For two years, she illegally stuffed and soldered thousands of printed circuit (PC) boards in her home. Her employer, a middle-aged woman she calls "Lady," subcontracted assembly work from big firms—so Maria was told—like Apple and Memorex.

Maria gladly accepted the low piece-rate work because child care would have eaten up most of her after-tax earnings at a full-time job. She quit, however, when Lady asked her to wash her assembled boards by dipping them into a panful of solvent, heated on her kitchen stove. Maria, unlike most Silicon Valley cottage workers, had studied chemistry before immigrating to the United States, and she knew that the hydrocarbon fumes could make her young son, crawling around on the kitchen floor, seriously ill.

Lady contracts with about a hundred minority women, primarily immigrants and refugees from Latin America, Korea, and Indochina. Although semiconductor chips are fabricated with precise machinery in super-clean rooms, they can be attached by hand, anywhere, to the printed circuit boards that form the heart of most computer equipment.

Silicon Valley is known worldwide for its engineers, scientists, and programmers; yet one in every twenty of its high-tech employees, or about ten thousand people, work for a printed circuit board subcontractor for little more than the minimum wage. The big-name electronics companies that send out PC-board assembly are familiar with the practices of their suppliers. Joe Weber, human resources manager of the American Electronics Association, told a reporter: "Many of the printed circuit board companies are pretty much sweat shops. . . . They pay what they have to to stay in business."

Most of the PC-board assemblers are minorities and women who cannot get better jobs. When the U.S. Immigration and Naturalization Service staged a series of raids in 1984, picking up and deporting undocumented immigrants, they found that as much as one quarter of the subcontractor workforce was in the United States illegally.

In addition, thousands of uncounted women like Maria assemble PC boards at home. They work at home either because they have young children, or because they cannot speak English, or fear deportation.

■ Silicon Valley's workforce is sharply stratified. In the electronics industry, pay, status, and responsibility are primarily a function of education. The professionals who make the Valley unique sit at the top of the occupational ladder; they are paid well, and the ambitious among them can make millions. Most are white men, but Japanese-Americans and ethnic Chinese are over-represented as well.

Skilled workers, such as technicians and craftspersons, are paid fairly well, despite the absence of union representation. More women and minorities work at such mid-level positions. However, skilled employees still make up a relatively small portion of the Valley's high-tech payroll, except at military electronics firms which do little mass production. Only rarely does a skilled employee move up into a professional position without earning a college degree. And it is difficult for an assembly worker to advance to technician or a craft job without outside training.

Although the semiconductor and computer companies do most of their labor-intensive assembly overseas or in other parts of the United States, semiskilled production workers still make up a third of the electronics workforce in Silicon Valley. Most of these semiskilled workers are women; about half are minorities—Mexicans and Chicanos, Filipinos, and Indochinese. To production workers, the Silicon Society means hazardous working conditions, expensive housing, long commutes, inadequate child care, and low pay. They live in a different world from their professional co-workers, although the two groups cross paths in company cafeterias and parking lots.

The world of Silicon Valley's managers and professionals is centered in northern Santa Clara County, near Stanford University and the historical center of the Valley's high-tech industry. Unlike the white-collar workers who commute to America's established downtown areas, Silicon Valley's affluent have chosen to live near their place of work. Other new, high-tech centers appear to be developing along a remarkably similar pattern.

Since Stanford University established its Industrial Park in 1951, high-tech companies have clustered near the university. The Industrial Park, on Stanford-owned land just a mile from the academic campus, established standards for industrial development in Silicon Valley, and it is still considered a model throughout North America. For three decades, its low-slung buildings, innovative architecture, and expanses of green landscape perpetuated the belief that high tech was a clean industry and a good neighbor. The suburbs around Stanford have long been known for their attractive living environment and good schools; and commuting, even before the 1973 rise in oil prices, was uncomfortable, costly, and time-consuming. So professional workers generally bought homes or rented as close to work as possible.

As the Valley boomed, its industrial core spread, but until the 1980s this core was for the most part confined to the northern, suburban portion. Like their predecessors, the engineers, scientists, and managers who came to the Valley from all over the world settled near their jobs. This influx of high-income families drove up the cost of housing. By the 1970s, rents and prices in the Valley were among the highest in the nation.

By and large, the unemployed, the service workers, and the Valley's low-paid production workers—who have always earned a fraction of the professionals' salaries—were driven from the centers of employment. San Jose, the county's traditional urban center and home to half its residents, became a bedroom community for the production workforce.

Palo Alto, which receives property and sales tax revenues from the Stanford Industrial Park, easily provides municipal services to its relatively affluent citizens. San Jose, on the other hand, has a much smaller tax base from which it must serve the county's poorer residents. Production workers from San Jose spend their days in the north county, generating wealth for electronics companies to pay into suburban treasuries. They then return to homes protected by San Jose's underfunded police and fire departments and streets maintained by its public works department.

Nowhere are the two worlds of Silicon Valley further apart than in education. Palo Alto's public school system is considered among the best in the nation; in fact, that is a major reason why high-tech professionals move to the area. In 1983, however, the San Jose Unified School District, the largest of several districts in the city, became the first American school system since 1943 to declare bankruptcy.

Most of the employed residents of East Palo Alto, the Valley's only black ghetto, work in Palo Alto just across San Francisquito Creek, but they live in another county. With no industry or luxury housing, the community has a tax base—measured as per capita assessed property valuation—less than one third that of Palo Alto. East

Palo Alto doesn't try to match Palo Alto's services; it has trouble just keeping its new city government in operation.

■ As Silicon Valley grew, the homeowners who dominate municipal politics in the north county cities strove to retain their suburban environment by restricting growth. In most cases, that has meant confining new residential construction to large, expensive homes. Industrial development on vacant land has continued unabated, and the demand for high-tech space has remained so high that in Mountain View, a twenty-year-old shopping mall has been converted into a sales and training center for Hewlett-Packard. But new housing construction, particularly for low- and moderate-income households, has been insignificant.

By the late 1970s, housing costs throughout the Valley were so high that government agencies recognized the problem. Even rents in San Jose, well below those in the north county, were rated the highest among American urban centers in the 1980 Census. With the cooperation of labor, industry, environmental, and other community groups, county officials launched a series of studies. They coined a phrase, "jobs–housing imbalance," to describe the shortage of housing in the Valley's traditional high-tech employment centers.

The problem wasn't new, but it had finally hit the Valley's richer half. Employers were finding it difficult to recruit out-of-town professionals, who could buy a mansion in Arizona or Michigan for the price of a small condo in Sunnyvale. The Valley's largest manufacturers formed the Santa Clara County Manufacturing Group to let the county and city governments know their concern.

The Manufacturing Group agreed with city and county planners that it was necessary to curtail industrial growth in the north county while expanding opportunities for residential development. But it did not put its influence behind rezoning. The manufacturers supported new land use policies in principle, but as individual firms they lobbied to retain their ability to expand on land near their existing facilities.

Nearly a decade after planners first identified the jobs–housing imbalance, the problem is worse. Some companies have set up shop in north San Jose, as well as in cities east of the San Francisco Bay, where they can draw workers from the bedroom communities of the East Bay. However, most new firms, as well as established ones, still prefer to set up or expand in the core of Silicon Valley, where support services are available and the bulk of the Valley's professional workers still live.

Outsiders hungry for their own share of high-tech development have predicted the demise of Silicon Valley. Indeed, the housing shortage is one reason why many established firms are carrying out some of their expansion elsewhere. So far, however, employees have adapted, usually unhappily, to the high cost of housing.

Groups of two or three families—particularly immigrants from Asia—are crowded into single-family houses and apartments. High-tech employees, including young professionals as well as production workers, move frequently to avoid the full impact of skyrocketing rents. Though tenants have qualified rent control initiatives for votes in several north county cities, each measure has been resoundingly defeated. Landlords have spent vast sums of money to convince home-owning voters that rent control would hurt their communities. East Palo Alto and San Jose, however, have enacted rent control.

Sometimes newcomers receive bonuses from their employers to purchase prime housing. But those who do not own their own homes are being forced either to move farther and farther from their jobs, or to dedicate huge shares of their income to housing. Many residents without ties to high tech, such as retired people, are simply leaving the area.

The long daily commute poses particular problems to the Valley's working mothers. Even when child care is available, parents find themselves uncomfortably far away from kids who may have problems or illnesses. Many school-age children are left home alone after school, and some preschoolers are left in the care of siblings not much older than they are.

■ In June 1984, a Superior Court judge temporarily blocked the construction of Regency Plaza, a proposed development of four thirteen-story office towers in an industrial area in the city of Santa Clara. Opponents of the project had filed suit, charging that Regency Plaza would generate twelve thousand car trips daily, clogging the area's already crowded roads. The developers soon bypassed the court case by obtaining new zoning from the city, and the plaintiff dropped the issue.

The plaintiff, in this case, was not an environmental organization or a group of homeowners. It was Intel, one of Silicon Valley's largest corporations. The semiconductor firm has a plant across the street from the Regency Plaza site. Senior Vice President Laurence Hootnick spoke to a reporter about the problem, complaining: "Try hiring employees that have to commute forty-five minutes to one hour."

Valley roads and freeways are so congested that most Silicon Valley employers are seriously concerned about getting their employees to and from work. "Rush hour" lasts three hours on a normal weekday. In spring 1984, viewers of the late-late-shows on Bay Area TV watched Peter Giles, the head of the Manufacturing Group, deliver a Free Speech Message on Bay Area Television warning that high-tech industry's human resources were being wasted in traffic jams.

The manufacturers demonstrated their political clout that November by pushing through a voter-approved tax increase in a state where tax relief is still a popular political goal. Santa Clara County voters supported a measure to add a half cent to the sales tax in the county for the sole purpose of widening, extending, or improving three freeways. Electronics companies and industrial developers put more than half a million dollars into the campaign. Evidently high-tech industry, which has consistently lobbied for lower corporate income tax rates and larger tax breaks for investors, had no qualms about favoring higher taxes when that meant getting the average resident to subsidize projects that high-tech companies consider vital to their growth.

■ A few years back, several women on the morning shift at Verbatim, a Silicon Valley manufacturer of memory disks for computers, complained of dizziness, shortness of breath, and weakness. Some even reported seeing a haze in the factory air. More than one hundred people were quickly evacuated from the building, and the company sent thirty-five of them to a nearby industrial clinic.

Hours later, inspectors from the California Occupational Safety and Health Administration could not find fumes intense enough to explain the complaints, and they termed the episode "mass psychogenic illness," also known as assembly-line hysteria. In the stressful world of high-volume electronics assembly, mass hysteria is not unknown. But chances are high that the Verbatim workers' bodies had detected the presence of toxic chemicals at a level below the threshold recognized by health officials.

High-tech industry's environmentally controlled "clean rooms," in which electronics workers must wear surgical gowns and gloves, are not designed to protect the workers; they are built to protect microelectronic products against particulate contamination. Despite the protective clothing, equipment, and vents found at a typical semiconductor plant, in the pressure to meet production quotas many Silicon Valley workers are frequently exposed to hazardous liquids and fumes.

The hazardous materials used in semiconductor production include acids, cyanide compounds, organic solvents, and silicon tetrachloride, which turns into hydrochloric acid when its fumes are inhaled into the lungs. Arsine gas, a lethal form of arsenic, can cause serious damage to the liver, heart, and blood cells even when inhaled in small quantities. It has been used extensively for years in the production of silicon chips. Now, as the Pentagon is promoting the development and production of chips based upon gallium arsenide instead of silicon, the likelihood of workers being exposed to arsenic is growing.

In order to help local workers protect themselves against such dangers, local activists in 1978 founded the Project on Health and Safety in Electronics, which became the Santa Clara Center for Occupational Safety and Health. In 1979, backed by funds from the Federal Occupational Safety and Health Administration, the Center established a health hazard hotline and produced a series of pamphlets on electronics industry workplace hazards.

High-tech management reacted immediately. Considering the Center a front for organized labor, semiconductor firms badgered safety officials to withdraw the group's grant. It took the election of Ronald Reagan to cut the federal purse strings. But the Center continues today, with a smaller staff, supported by organized labor, foundations, and church organizations.

Although the dangers of working inside a high-tech plant have not received the extensive press coverage afforded to leaks and spills of toxic chemicals into the Valley environment, there is a growing recognition among health professionals that high-tech production creates its share of acute illnesses, and further, that high-tech workers face an increased risk of cancer, liver disease, and other chronic problems as a result of their exposure to toxic chemicals.

Some companies have relatively good health and safety records, but the semiconductor industry as a whole has tried to "kill the messenger" bearing bad news rather than attempt to clean up its act. In 1981, semiconductor companies unilaterally redefined "one-time exposures" to toxic materials so as to improve the industry's official occupational safety record. By classifying exposures as "injuries," not "illnesses," they reduced their historically high reported incidence of occupational illness with the stroke of a pen. Under California law, injuries need not be reported if no work time is lost, but illnesses must be reported in any case.

■ High-tech production is unsafe, and its workers are underpaid, yet high-tech workers have not rushed to follow the union banner. Very few are represented by labor organizations nationally, and in Silicon Valley, not one computer or semiconductor firm has a recognized union bargaining unit. The last publicized organizing drive, held at Atari, faltered in November 1983 when the Glazier's Union lost a representation election by a vote of 143 to 29. To be successful, union organizers in the Valley have to overcome the labor movement's negative national image, and they need sufficient resources to combat the ploys of determinedly anti-union managers.

However, the chief reason that unions have not yet cracked high tech is that it is a new industry. The textile, auto, and steel industries were not organized in a day; it took time to lay the organizational foundations for unionization. More important, as a new, rapidly changing and growing industry, microelectronics is characterized by a high rate of employee turnover. When workers are dissatisfied, they may sign a union card. But the most practical action, from an individual point of view, is to get another job. Since Silicon Valley industry is expanding, that usually is not too difficult. By the time a bargaining election comes around, the unhappiest workers are usually gone.

Unions have also been hamstrung by their inability to win the loyalty of Asian immigrants. Filipinos and Indochinese make up a significant portion of the Valley production workforce, and most of them appear to harbor little sympathy for organized labor. This is not an ethnic characteristic, however, as the strong working-class movements in Indochina and the Philippines reveal. Rather, U.S. immigration and refugee policy has largely confined Asian immigration to anti-Communist Vietnamese and upper-middle-class Filipinos who were qualified, in the Philippines, for professional careers. In the United States, many Filipino women work in electronics while they study to qualify for nursing licenses and teaching credentials.

Employers and union organizers agree that Mexican and other Hispanic immigrants, whose entry into the

United States is not effectively controlled by the Border Patrol, tend to be strong supporters of the labor movement here, but there have never been enough of them at any one plant to vote in a union. Blacks, who tend to favor unionization because they have historically benefited from it, make up only about 5 percent of the Valley workforce.

Often workers are more dissatisfied with conditions that they do not perceive to be work-related—such as expensive housing, long, slow commutes, and the absence of affordable child care—than with their employers. Better pay would help to solve those problems. However, a worker who receives a $100 a month raise is unlikely to blame her employer for her low pay, even if the added income does not cover her increased cost of living. It is easier to blame the landlord who hikes the rent by $100 per month.

To organize Silicon Valley, unions need to fund and support a long-term, industry-wide organizing drive. Such a drive would follow workers from employer to employer, and it would link "community" issues to workplace organizing. It would also back the formation of ethnic committees, to be headed by workers and organizers from each culture present in the workplace.

But a drive of this sort, which would not bring immediate results in the form of union dues collected by the employer, would require the investment of substantial "risk capital" by the labor movement. Labor officials familiar with the Valley favor such an approach, yet those international unions that might be expected to back the drive—the Auto Workers, Communications Workers, Electrical Workers, Machinists, etc.—are on the defensive, since they are already losing members to automation and new competition. Proponents of high-tech organizing argue that unions must organize growing fields like high tech to compensate for their losses in other industries due to the application of microprocessors to the workplace. But those labor leaders with their hands on the purse strings fear that a costly drive in Silicon Valley would fail.

The Valley employers are very competitive manufacturers, who have a constant interest in lowering production costs. As producers of high-tech equipment, they are familiar with the latest techniques in industrial automation. When their product lines mature, they do not hesitate to replace people with machines.

As the price of automating production falls, the value of semiskilled labor falls as well. Unless electronics workers organize, their economic condition can only get worse. Living conditions may be bad now, but in a few years they may become untenable. Violence, in the form of random attacks or even riots, could erupt as a result.

■ Linda Cartlidge is a Silicon Valley success story. Her company, Cartlidge and Associates, stages elaborate trade shows like the Advanced Semiconductor Equipment Expo and the Word Processing and Office Equipment Expo. She has numerous full-time employees, and during shows she hires up to two hundred men and women.

"Microelectronics is one of the few areas where a woman has a chance to advance very quickly," Cartlidge comments. "Any time you get into a new and burgeoning environment, there's a chance for you not only to get in on the ground floor, but to meet the type of people who can help you advance." In Silicon Valley, many educated women like Cartlidge have worked their way up from clerical jobs into positions in management.

In most cases, top management is still the preserve of

male engineers, financiers, and marketing specialists from other industries, but women are found in positions of responsibility in personnel, marketing, publications, and public relations. Like the hard-driving engineers who leave their employers to start up new, spin-off companies, ambitious female professionals have established high-tech service businesses.

However, in the workaholic world of Silicon Valley, women with children cannot succeed without sacrificing their family life. Cartlidge, a single parent, says she and her son Cord often went hungry when she worked as a secretary. After she moved up into management, keeping long hours, she put Cord into a boarding school.

Getting a new product out ahead of the company down the street is frequently essential for corporate survival, so professionals and managers in Silicon Valley are expected to work long hours. Members of Apple Computer's Macintosh development team, for example, wore T-Shirts emblazoned WORKING 90 HOURS A WEEK AND LOVING EVERY MINUTE OF IT. In fact, the high stress and long hours built into high-tech's industrial culture threaten the home life of all Valley professionals with families. Psychologists who have studied the "silicon syndrome," the devotion of professionals—usually men—to their jobs in electronics, blame the Valley's high divorce rate on the syndrome. Engineers and executives who retain their dedication to the family often find that others have leapfrogged them on the ladder to success.

For mothers like Cartlidge, the choice is particularly painful. To keep up with their male colleagues, they must put in long hours on the job. In most cases, successful Valley women report that this means their children spend a lot of time alone. Frequently deals are made and ideas are exchanged in bars or restaurants after hours, but working mothers—unless their husbands are unusually liberated—must go home to prepare or serve dinner. Even when child care is available, the mother is usually the one to deliver the child in the morning and pick him up in the afternoon. The fabled recreational facilities at companies like Rolm, which are designed to help relieve work stress, are open to working women. But mothers of young children cannot afford to stay late.

■ Despite all its problems, Silicon Valley is still the number one address in high tech. The roads are congested and the cost of living is exorbitant, yet the Valley's lifestyle still attracts professionals from all over the world. Large, established firms are building some of their new plants elsewhere, but years ago the Valley achieved a critical mass of innovative energy which other areas may never match.

According to popular Valley mythology, the rugged individualism of entrepreneurs and venture capitalists is responsible for its success. However, the U.S. government, in the form of Pentagon research and procurement contracts, created the conditions that made Santa Clara County a high-tech Mecca. Military contractors dominated the area's high-tech industrial complex until the 1970s, when firms with a commercial focus took the lead in sales, employment, and technology.

Today, the Pentagon wants to reassert its hegemony over high tech. It is funding massive new research programs like Strategic Computing and the Very High Speed Integrated Circuit program, and it is attempting to control the flow of high-tech ideas and the trade of high-tech products.

The subtle battle between civilian high tech and the military industrial complex has political overtones, but it is essentially a fight between two approaches to technology. The creative, independent young scientists and engineers who put Silicon Valley on the map are unlikely to play the Pentagon's game.

■ In 1938, Stanford electrical engineering professor Frederick Terman helped two of his students, William Hewlett and David Packard, to build audio oscillators in a Palo Alto garage. Their company, Hewlett-Packard, made its first sale to Walt Disney's studios for the production of *Fantasia;* today it is the Valley's largest home-grown firm. To Terman, Hewlett-Packard served as model for a much larger high-tech partnership between university and industry, which he called the "community of technical scholars."

In an area where industrial leaders are known more for gambling on new ideas than for their long-term vision and planning, Terman stands out as a remarkable man. He not only saw the future but had the resources to make it happen. A product of his times, he did not hesitate to seed his vision with university research contracts from the Department of Defense.

After directing Harvard University's Radio Research Laboratory during World War II, Terman returned to Stanford to serve as dean of engineering and later university provost. He built up Stanford's electrical engineering department with military research and construction contracts, and he helped professors and graduates to establish nearby companies, such as Watkins-Johnson and Applied Technology, to enable them to apply their research. He also convinced local electronics companies to donate funds to the engineering school, and he welcomed their employees as both students and instructors. Soon the Stanford area boasted a world-renowned pool of technical brainpower.

It was Terman who persuaded the university's trustees to establish the Stanford Industrial Park on vacant university land. The park not only welcomed Stanford spin-offs but acted as a magnet, attracting established high-tech firms from outside the area. For example, in 1956 Lockheed, then primarily a manufacturer of aircraft, sought land for the research arm of its new Missiles and Space division. It chose the Stanford park to take advantage of Terman's growing community of technical scholars, and it liked the area so much that in 1957 it established its much larger manufacturing facility in nearby Sunnyvale. Soon the Lockheed Missiles and Space Company was the largest employer in the Valley; it remains so today, with about 23,000 on staff.

Lockheed, Philco (now Ford Aerospace), Sylvania (GTE), and other large employers relied almost exclusively on military and NASA contracts for their revenue. Fairchild Semiconductor, the Valley's first successful semiconductor firm, received large orders from the space program at a time when integrated circuits were too expensive for widespread commercial use. Even Hewlett-Packard, which manufactured instruments, not weapons systems, depended heavily on the government and government contractors for its sales.

During the late 1960s and early 1970s, however, Silicon Valley outgrew its military roots. In the wake of anti-Vietnam War demonstrations at and around Stanford University, where activists challenged the local military electronics effort, a growing number of young technical professionals sought out nonmilitary work. Even pro-military scientists and engineers found that working for the military or its contractors stifled creativity. High-flying new companies, such as Intel and System Industries, eschewed Pentagon business to avoid the paperwork.

Antagonism toward the military and other large organizations actually provided the initial impetus for one of the best known and fastest growing segments of the high-tech industry, the manufacture of personal computers. The first microcomputers, data-processing machines based upon one microprocessor, were created by hobbyists who not only shied away from the military's workstyle but actively opposed its goals. Fred Moore, the man who called the first meeting of Palo Alto's "Home Brew" computer club in 1975, was and still is a draft resistance organizer. Another club member, Lee Felsenstein, moved from designing and building home-brew bullhorns for anti-war demonstrations to designing the Sol and later the Osborne personal computer. Apple founders Steve Jobs and Steve Wozniak, also members of Home Brew, were not active in peace organizations, but they were sympathetic. Only after companies like Apple and Osborne had proven successful did the big guns in the information industry, such as IBM and AT&T, enter the personal computer sweepstakes.

Although the Pentagon pours $4 billion directly into the Silicon Valley economy each year, the military has lost its hegemony. Many of the brightest young technical scholars, given the choice of working in the more open "culture" of the civilian sector, reject the stodgy environment of military contractors like Lockheed. Despite the high volume of secret military work in the area, Silicon Valley's burgeoning commercial sector has created an intellectual climate conducive to technological innovation.

■ In virtually every region of North America, politicians or business leaders see their community as the next Silicon Valley. Each state has developed incentive programs, ranging from training programs to capital grants, to lure investment by high-tech entrepreneurs and venture capitalists. But even under the best of economic conditions, there will never be enough high-tech production to go around.

There is a risk, therefore, that development officials will essentially give away the store to attract high-tech companies. Texas, for instance, promised more than $60 million in educational programs and construction assistance to win a fierce bidding war between the states, enticing the new Microelectronics and Computer Technology consortium to set up shop in Austin. Oregon has dropped the unitary method of calculating state income taxes, foregoing revenue to attract Japanese high-tech manufacturers to the state.

But any local or state government welcoming the electronics industry must be prepared to deal with the problems of high-technology development. If they do not have the funds to plan adequate housing, build roads and sewage plants, regulate workplace hazards and environmental pollution, encourage child care, and enforce laws against illegal employment practices and white collar crime, then residents will bear the costs of the Silicon Society directly.

Silicon Valley achieved its critical mass of technical expertise long before its quality of life deteriorated. Yet if those communities that are just now starting to build high-tech complexes fail to plan to accommodate growth, they may lose their appeal, compared to other high-tech centers, and never realize their dreams of a Silicon Rush.

Steve Jobs, who is now a thirty-year-old multimillionaire, told visiting French President François Mitterand that the Valley's unmatched rate of industrial innovation is the result of its developed supply of venture capital. Investors willing to risk their millions on new products have made it possible for high-tech entrepreneurs to try their luck in the marketplace. Jobs and his friend Steve Wozniak, for instance, could not interest their employers, Atari and Hewlett-Packard, in their idea for mass-producing microcomputers. But with backing from venture capitalists, they launched Apple as a dominant force in the new microcomputer industry. When Apple placed its stock on the market, of course, Jobs, Wozniak, and their investors struck gold.

In order to attract competent professional staff, start-up firms in the Valley offer lucrative stock options. Unable to pay high salaries at first, they lure engineers and managers with the opportunity to buy stock at low initial values. If the company succeeds, these employees also can sell stock and strike it rich.

When it works, venture capitalism appears to be an excellent mechanism for stimulating the development of new technologies. But too much venture capital can undermine the health of high-tech companies. In 1983 and early 1984, investors flooded the Valley with high-tech funds. Established firms like Intel had difficulty hanging on to their key employees, since venture capitalists were willing to back virtually anyone with experience and ambition.

Some Valley executives are worried that careless entrepreneurial activity might kill the goose that laid the silicon egg. A top Xerox executive suggests that many in the latest generation of entrepreneurs do not intend to follow the lead of Apple's Jobs or Intel's Robert Noyce to create viable ongoing enterprises. Instead, they would appear content to emulate Adam Osborne, who started Osborne Computer, or Chuck Peddle, the founder of Victor. Both these men struck it rich in the computer business, but their companies, located on the outer edges of Silicon Valley, went bankrupt.

The promise of personal wealth intrinsic in venture capitalism has a dark side, even when new high-tech firms are successful. As the gap between the Valley's rich and poor widens, critics wonder whether it is possible to generate an industrial environment like Silicon Valley, where new products, processes, technologies, and companies are the norm, without creating a sharply skewed distribution of income and wealth.

In the global rush to emulate Silicon Valley, no one has seriously sought an alternative form of financing. However, it is surely possible to create sources of capital with which to gamble on new products and technologies without offering vast riches to the founders of high-tech companies. Even today the average Silicon Valley professional produces creatively and diligently for much more modest rewards. Not surprisingly, those who have made millions argue, like Jobs, that wealth is the primary motivating factor. If we are to build a Silicon Society as remarkable as the technology upon which it is based, then we must find ways to spread the rewards.

If, on the other hand, the present trends continue, the two worlds of Silicon Valley will drift farther apart. And the gap between them is potentially explosive.

Chapter Eight
THE TOXIC TIME BOMB

■ ■ ■ ■ ■

Early in 1984, a producer of printed circuit boards for computer and other high-tech manufacturers announced that it was closing its Mountain View plant, in the heart of Silicon Valley, because it could not comply with local ordinances regulating the storage and disposal of hazardous materials. CTS Printex, which employed more than four hundred people, said that it was moving across the San Francisco Bay to Fremont, on the eastern edge of the Valley's ever-expanding electronics complex.

The announcement provoked a protest, but not in Mountain View. No one there questioned the need to enforce the community's increasingly strict environmental ordinances. In Fremont, however, neighbors of the proposed plant joined with a local group called Sensible Citizens Reacting Against Hazardous Materials to tell city planners flatly that they did not want a firm with a dirty track record operating in their neighborhood.

Fremont and Mountain View residents, along with anybody else who reads a newspaper or watches TV news in Silicon Valley, are all too familiar with what still ranks nationally as one of high-tech industry's greatest secrets: Electronics is a dirty industry. It is possible that communities and regions which study the lessons of Silicon Valley can substantially reduce the risk high-tech production poses to the environment and public health.

Unfortunately, high tech's environmental record has not leaked out to the rest of the country. Officials who promote high tech as a solution to local or regional economic ills paint a picture of the industry as shiny as the surface of a silicon wafer. They call high tech a "sunrise industry," clean and light in contrast to "smokestack" industries like steel and auto production, known for their drab, monstrous factories and ever present plumes of vapor and smoke. In June 1983, for instance, Texas Governor Mark White, having lured a major new high-tech venture into his state, told reporters: "I don't think you'll find that there will be any pollution [from the electronics industry] unless the Japanese cars they drive to and from work do it."

It isn't hard to see where high tech got its reputation. Electronic products—chips, computers, switchboards, and so on—don't breathe exhaust or drip oil. The factories are rambling, well-landscaped buildings, resembling modern college libraries; no smokestacks protrude above their facades. Many production steps take place in so-called clean rooms, where the air is fanatically filtered and production workers wear surgical gowns. But the industry's vast investment in cleanliness is designed principally to protect microelectronic components from the dust particles that could prevent them from functioning properly. It does not protect high-tech's workers, nor the residents who live in the communities that surround the plants, from the toxic chemicals and metals essential to high-tech manufacturing.

One of the greatest ironies of microelectronics technology is that the transformation of America into an information society relies, at its core, upon a technology from the industrial era: chemical processing. The manufacture of chips, printed circuit boards, magnetic media, and other high-tech products uses some of the most dangerous materials known to humanity. And the accidental release of those toxins into the air, the ground, and bodies of water poses a significant threat to public health.

High-tech pollution is a fact of life wherever the industry has operated for any length of time, from Malaysia to Massachusetts. Yet nowhere has the growing threat that electronics production poses to public health been clearer than in Silicon Valley, where the concentration of high-tech production has greatly magnified the industry's environmental problems.

The hazards of high tech have become increasingly clear during the past few years, but it may be decades before the full impact on public health is known. The electronics industry uses thousands of different toxic materials, yet the volume is small compared to chemical-intensive industries such as petroleum and pesticide production. Still, a Bhopal-like incident, in which hundreds of people are killed immediately from a single leak, is a serious possibility.

Even without such a catastrophic accident, however, the long-term toll from high-tech pollution may be enormous. High-tech toxics have been slowly entering the environment of Silicon Valley for decades. Though widely used chemicals such as hydrocarbon solvents are known to cause ailments ranging from headaches and birth defects to cancer, it is difficult to demonstrate that any particular person is a victim of a particular leak or spill. But there is no doubt that industrial chemicals are affecting the health of growing numbers of people.

■ San Jose attorney Amanda Hawes is one of a handful of Silicon Valley activists who warned for years that high tech was indeed a hazardous industry. She has built up her reputation by representing electronics workers injured by chemicals on the job. Today she also represents residents of the Los Paseos neighborhood in southern San Jose. A new, comfortable, working-class suburb typical of Silicon Valley, Los Paseos is distinguished by the presence of a chip manufacturing factory built by Fairchild Semiconductor in 1975.

Hawes carries with her a large zoning map of the area surrounding the Fairchild plant. On every block in the surrounding neighborhood there are several colored pins and flags. Each triangular red flag represents a child born with heart anomalies; each blue pin marks a miscarriage; each yellow flag signals a cancer case. Black flags, superimposed on the other markers, note recent deaths. Hawes also carries with her a supply of pins, and she must frequently add one to the display. She charges that Fairchild is responsible for the area's high incidence of disease.

Most of Hawes's clients believed that electronics was a pollution-free industry until January 1982. At that time, officials disclosed that six weeks earlier they had shut down a drinking water well operated by the Great Oaks Water Company, just 2,000 feet from an underground chemical storage tank at Fairchild. Solvents from the tank, including suspected carcinogens trichloroethane and dichloroethylene, had entered the water supply. When residents learned of the leak, they quickly concluded that the company was to blame for the area's alarmingly high incidence of birth defects and miscarriages.

Since then, Fairchild has spent at least $15 million to reduce the concentration of solvents in the aquifer, but the water will never be as clean as it was before Fairchild set up shop there. Now the factory stands empty, a monument to the dying myth of high tech as a clean, light industry.

The Fairchild leak exploded onto the local front pages and six o'clock news, breaking through a long-standing barrier of silence on high-tech pollution. The Bay Area press, public officials, and electronics corporations themselves have all been forced to investigate environmental hazards that nobody wanted to believe existed.

Today, scarcely a week passes without the revelation of a new leaking storage tank, poisoned well, or pollution law violation. As soon as the extent of the Fairchild leak was known, other companies started to test the ground water around their underground chemical tanks, and the Bay Area's Regional Water Quality Control Board ordered a comprehensive testing program. Most of the Valley's large production sites were checked—and most came up dirty. Even firms with a reputation for environmental concern, like Hewlett-Packard, had been leaking dangerous toxics used in their manufacturing processes.

Leaks were found at scores of industrial locations within Santa Clara County, but many small facilities have still not been tested. Nineteen high-tech sites have been placed on the Environmental Protection Agency's "Superfund" list. Nine public and more than sixty private wells have been shut down; many others contain legally allowed levels of contamination. Luckily, Silicon

Valley residents have thus far been spared an outright environmental disaster. The Valley's largest source of drinking water is protected by a 200-foot layer of clay, which separates polluted ground water from deep aquifers.

Though Fairchild and nearby IBM began the task of clean-up soon after pollution from their facilities was discovered, many Valley electronics firms have not done much more than sink test wells to determine the extent of their leaks. Pools of hazardous chemicals drift around underground, poisoning shallow private wells and possibly finding a route—for example, via an abandoned agricultural well—to the public water supply. Unless the toxic chemicals are removed or neutralized before they percolate through the clay, the primary water supply of several hundred thousand people will be permanently poisoned. Silicon Valley is sitting on a toxic time bomb. No one knows when it is set to go off; certainly, not enough is being done to defuse it.

Despite the slow clean-up rate, Silicon Valley's governmental agencies and high-tech companies acted quickly to develop rules and storage procedures for preventing future leaks. Their approach, the core of which now is built into both federal and California state law, is serving as a model for the regulation of hazardous materials, including gasoline as well as high-tech toxics, across the nation.

Soon after the Fairchild disclosure, fire chiefs representing Santa Clara County's individual cities and fire districts established a task force to develop tighter local regulations governing the handling and storage of hazardous industrial materials. The task force was asked to develop a model ordinance for passage by each industrial city in the county plus the county government, which governs unincorporated areas. The fire chiefs chose not to look at occupational health and safety issues, which are currently regulated by federal and state agencies; they focused instead solely on chemical leaks and potential fire hazards. They sought to close a major gap in existing regulations by establishing standards for the storage of nonflammable, "virgin"—that is, awaiting use—industrial toxins.

Almost immediately, trade associations representing the Valley's high-tech companies asked to participate in the task force. Welcomed by the chiefs, they sent engineers, not PR flacks. Rather than oppose the ordinance outright, the engineers sought to shape a regulatory program with which they could live. Industry representatives agreed to a requirement that they install double-walled containers for all new underground toxin storage. The principle is both simple and sound: If an inner tank leaks, the secondary barrier will contain the chemical or waste. This simple standard should prevent most future leaks from entering the environment.

In return, the fire chiefs accepted industry's argument that it would be impractical to replace all existing tanks, some of which were built under the factory floors. Under the proposal, old single-walled tanks could remain in use as long as a rigorous monitoring program, also required by the ordinance, detected no leaks.

A few months after it began its work, the task force held its first public hearing. Men in business suits packed the Sunnyvale city council chambers to hear initial reports from the group's committees. Though open to the public, the hearing had not been advertised. The dialogue, it appeared, was planned solely between city officials and high tech.

Community, labor, and environmental groups learned of the meeting through the newsletter of an industry association, and a handful of representatives from public-interest groups testified. Soon they formed the Toxics Coalition, which then injected itself into the deliberations, forcing the adoption of strong public disclosure and "whistleblower" protection sections. Despite the lack of union representation in the electronics industry, organized labor mobilized the activists—from firefighters', construction, aerospace, and other unions—that gave the Coalition its clout.

Though there was never any doubt that Silicon Valley's environmentally sensitive communities would enact ordinances to regulate toxic storage, high-tech industry's official leadership took the unusual step of lobbying for strong, uniform regulation. In the city of Mountain View, for instance, a public hearing began as a caricature of the national conflict between environmentalists and business. A small gas station owner decried unnecessary government intervention, while a spokesman from GTE Sylvania asked the city to enact weaker rules than those proposed by the fire chiefs' task force. Members of the Toxics Coalition urged strong regulation.

Then a member of the audience rose to back the environmentalist position emphatically. There was nothing original about his presentation, but his position was unique. Larry Borgman, manager of plant facilities at Intel, speaking as the official representative of Silicon Valley's four major trade associations, offered industry's unequivocal support for the legislation. GTE's executives never spoke out against the measure again.

The model ordinance, written to apply to the underground storage of all toxic chemicals, drew opposition from the petroleum industry. At first service station owners and major oil companies objected to the proposal; then they argued for exemption. Finally, they tried to weaken the standards. Unlike the electronics industry, "big oil" sent its PR men to discuss the problem.

Electronics executives suggested that all industries

should be treated equally, and the Toxics Coalition reminded officials of the numerous underground petroleum leaks discovered in the area. After all, commercial gasoline contains 2 to 4 percent benzene, a recognized carcinogen, as well as lead and other toxins. In the end, the Santa Clara County Inter-Governmental Council, representing the cities, county, and other public agencies, recommended a strong model ordinance; most of the county's cities enacted its provisions and budgeted funds for its enforcement.

Since then, both the state of California and the federal government have enacted legislation imposing double containment standards similar to those developed in Silicon Valley. California has already registered over 114,000 existing underground chemical storage tanks.

So, some progress has been made. Without enforcement and citizen awareness, however, laws on the books will have no effect on the potential toxic time bomb that high-technology manufacturing represents. It will take public pressure, over several years, to force the state and the Environmental Protection Agency to fund adequate levels of enforcement.

■ In the wake of the Fairchild tragedy, leaks from underground storage tanks have received the greatest public attention in the Valley; but sewage from high-tech plants also threatens the environment. Despite federal and state regulation, concentrations of toxic heavy metals such as nickel, lead, and cadmium are building up in the San Francisco Bay, entering the food chain through the Bay's surviving fish population.

The manufacture of semiconductors and printed circuit boards creates large quantities of liquid waste. Producers are required to pretreat their toxic waste, precipitating it into a sludge for disposal at approved landfills. Sewage treatment districts, worried that toxic effluent could damage the organisms that process the sewage at their treatment plants, closely monitor industrial effluent for its heavy metal and toxic chemical content.

In 1983, Citizens for a Better Environment (CBE), a San Francisco–based watchdog group, reviewed the records of sewage treatment agencies and found widespread violation of pretreatment standards throughout the Bay Area, including Silicon Valley. Those records, of course, did not list the scores, perhaps hundreds of small electronics firms that routinely pour hazardous materials directly down the drain. To compound this, the problem may grow more severe. As the cost of proper treatment rises, the temptation to dump industrial chemicals illegally rises.

In response to the CBE study, regulators are now cracking down on at least some of the violators. In February 1984, for instance, operators of the San Jose sewage plant, which serves Santa Clara, San Jose, and several smaller communities, threatened to revoke the discharge privileges of five printed circuit firms. In July 1984, the Environmental Protection Agency ordered thirty-two Valley firms to pretreat their effluent properly or face fines of up to $10,000 per day.

Inspectors say that most of the Valley's large manufacturers and many small ones comply with existing pretreatment standards. Such rules, however, may not be enough to protect the Bay, which is slowly being poisoned by high-tech sewage. As Mike Belliveau of CBE warns: "Those standards originally were drafted more than ten years ago and were based on deep ocean discharge, not for a dead-end slough at the south end of the Bay where the water is only six feet deep."

The development of new industrial processes that will generate less waste is necessary if the growing environmental load of high tech is to be contained. Pretreatment is reaching its limits, since materials discharged in extremely small amounts can rise to high concentrations in the treatment process. The presence of gold in Silicon Valley sewage, while not considered an environmental hazard, illustrates the problem. Many electronics firms use gold to build their products, and no manufacturer, large or small, is going to be lax in the discharge of gold into sewage lines. Yet the Palo Alto sewage plant sells its sludge to a firm that burns the sludge and "mines" the ash for gold!

■ The sludge generated in sewage pretreatment, as well as other bulk chemicals and hazardous solid wastes, also pose a serious environmental hazard. Since approved Class I dumpsites—those specifically designed to handle hazardous wastes—are located far from the communities which offer a quality of life attractive to high-tech firms and professionals, tons of hazardous materials are trucked along the highways every day. Accidents are inevitable.

In late 1981, for instance, a tank truck carrying wastes from a Silicon Valley electronics firm pulled over at a roadblock in San Ramon, on the eastern side of the San Francisco Bay. Inspectors were looking for embargoed fruit, but instead they found a potential chemical disaster. The truck was leaking. Four thousand nearby residents and schoolchildren were quickly evacuated, yet at least twenty-two people had to be treated at local hospitals for respiratory complaints. The truck had been transporting copper, zinc, chromium, and titanium wastes, as well as sulfuric, nitric, hydrochloric, and acetic acids, to a Class I hazardous waste site.

Nobody knows how much illegal high-technology waste dumping actually occurs. In June 1983, officials in

Southern California found nearly eighty 55-gallon drums of hazardous chemicals in a stolen trailer in a parking lot in the remote community of Castaic. Investigators traced the wastes to Silicon Valley's Electrofusion, which had hired the Silicon Valley Hazardous Waste Transportation Company to haul away its hazardous wastes.

Electronics companies are not held responsible for the actions of contractors that are licensed to haul hazardous materials, but they are generating new forms of hazards faster than the waste handlers can update their methods. There are state and federal regulations to govern the waste transportation and disposal industry, but the enforcement agencies are hampered by being both underfunded and understaffed.

Furthermore, even when they are handled in accordance with the latest technologies and regulations, hazardous wastes represent another toxic time bomb. Many chemicals generated by the electronics industry, such as the hydroxides formed in sewage pretreatment, are likely to outlast the dumpsites at which they are "disposed." In October 1984, the Environmental Protection Agency fined the operator of the two dumpsites closest to Silicon Valley $161,000 for leaks.

In June 1985 it sought $7.4 million in penalties from another disposal firm, while six environmental groups charged that all dumps receiving Silicon Valley's hazardous wastes were leaking.

Even if disposal companies meet EPA standards, they are holding materials which, like nuclear waste, retain their toxicity almost indefinitely. Disposal in geologically suitable sites may be safe in the short run, but future generations will have to face the consequences of dangerous leaks unless better disposal or reprocessing techniques can be developed.

Both the state and federal governments are slowly enacting regulations restricting the disposal of particularly dangerous classes of toxic wastes; however, the rules contain loopholes allowing for land disposal until viable alternatives are found. As long as the loopholes are there, it is unlikely that either the electronics industry or the petrochemical industry will develop and test large-scale alternatives.

■ If any one visual characteristic has distinguished high-tech factories from the behemoths of heavy industry, it has been the absence of smokestacks. In Silicon Valley, municipal architectural standards require that manufacturers hide their vent outlets behind fancy façades. Nevertheless, high-tech plants spew tons of smog-producing fumes and toxic gases into the air every day. More dangerous still, a minor industrial accident, such as the rupture of a single cylinder of a commonly used gas like arsine or phosphine, could hospitalize or even kill scores or even hundreds of local residents.

For an industry that is reputed to be "clean and light," the scale of pollution is phenomenal. The Bay Area Air Quality Management District estimates that semiconductor firms alone in Silicon Valley emit 10 tons per day of "ozone precursors"—hydrocarbon solvent vapors which are chemically transformed into smog. Eight corporations are reportedly responsible for 74 percent of that total. In June 1983, the District enacted new controls on atmospheric emissions from semiconductor plants, designed to reduce vapors by a total of at least 3 tons per day.

High-tech executives treated the proposed air-quality standards in much the same way as they handled the storage ordinance. They did not oppose the regulations outright. Instead, they worked closely with the District staff to win technical modifications and delays that would ease the cost of compliance.

Much to the chagrin of the Toxics Coalition, however, the Air Quality District has delayed consideration of rules governing "exotic" emissions. These are chemicals, including potentially lethal gases such as arsine and phosphine, which are vented in small quantities by high-tech firms. In low concentrations, toxic gases might affect the health of only one in ten thousand, or one in a hundred thousand people. When several hundred thousand people are exposed over a period of time, however, the danger is real. Assessing the risk is a complex matter, so the District is waiting for California's statewide Air Resources Board to measure and establish standards for toxic contaminants. The process could take years.

In the meantime, just as they have already fouled the land and water in their communities, high-technology manufacturers continue to pour invisible poisonous gases into the atmosphere.

■ Silicon Valley's mounting experience with high-tech pollution provides lessons for the literally hundreds of North American cities that hope to become new Silicon Valleys. Any community which hopes to share the benefits of electronics development must be prepared to pay the environmental or regulatory costs. Those communities which, desperate for jobs, relax their environmental standards, are making a terrible mistake. Not only are they risking environmental disaster, but they are unlikely to bring in additional investment. Most high-tech companies don't look closely at such regulations when making siting decisions.

In particular, would-be Silicon Valleys need to prepare regulations covering the potential hazards of high tech before they invite the industry into their area. And they must develop the ability to enforce those rules competently before high-tech companies are ready to commence production. Communities are better protected and managers are happier when pollution control devices are built into plants from the ground up, rather than retrofitted by an arbitrary deadline. The engineers who manage electronics companies prefer dealing with public agencies that know precisely what they are controlling; they dislike working with regulators who act sympathetic but offer only vague guidelines.

In fact, some industry executives look favorably upon environmental protection. High-tech firms are more dependent upon brainpower than chemicals, and they must recruit professional workers from a worldwide job market. Silicon Valley is the world's leading center for high-tech industry, largely because it offers a quality of life which attracts these professionals. But if its water supply becomes permanently poisoned, few young scientists and engineers—especially those with children—will want to migrate to the Valley. This need to recruit skilled workers, more than any other factor, explains the industry's uniquely cooperative relationship with its regulators. How can an employer attract a key engineer from New York or a highly trained programmer from Oregon if the recruit knows that the local drinking water contains dangerous levels of industrial poison?

In the San Francisco Bay Area, where the environmental hazards of high tech are now well known, some residents remain dissatisfied with technical means of pollution control, such as double-walled containers, scrubbers, and pretreatment facilities. They have sought to limit their exposure by keeping high-tech production at a distance. Opponents of a Hewlett-Packard project in Rohnert Park, Sonoma County—a two-hour drive north of Silicon Valley—petitioned for a referendum that would have blocked construction. They lost the vote in April 1982, in part because Hewlett-Packard has one of the Silicon Valley's best reputations; but the company was forced to scale down its development plans to win approval.

In Silicon Valley itself, when the owners of a shopping mall announced in March 1983 that they intended to convert the entire complex to light industry, neighboring homeowners organized in opposition. The conversion was approved, but only on the condition that toxic use be reviewed by independent consultants and government. The neighbors still initiated a petition drive against the project; they dropped their campaign when Hewlett-Packard announced plans to lease the whole shopping center as a sales and training center, with no manufacturing.

When high tech does move into an area, it is also important to get regulatory agencies to work together. In Silicon Valley, both regulators and companies have been hobbled in their attempts to solve hazardous materials handling and disposal problems by the network of municipal, regional, state, and federal regulations. High-tech firms installed storage tanks underground to comply with fire regulations. They concealed vents to comply with municipal architectural standards, thus forcing toxic fumes back into facilities through the air-conditioning intakes. However, vapor scrubbers, which were installed to reduce air pollution, add to water pollution. And the processes designed to remove heavy metal wastes from liquid sewage precipitate them into much more stable chemical forms, requiring permanent land disposal. A study of Silicon Valley's environmental problems now being conducted by the EPA's Integrated Environmental Management Project should highlight the need for consistent regulation.

However, the future of effective protection depends primarily upon the willingness of authorities to work together.

In the long run, the solution to the high-tech toxic time bomb lies not in controlling pollution but in developing production techniques that will reduce the use of hazardous materials. Public regulatory policy should encourage the adoption of new, more environmentally sensitive production methods by making hazardous approaches more costly. For example, when the Air Quality Management District established tighter "photoresist" process emission standards for Silicon Valley chip makers, it offered two methods which companies could use to achieve the District's goals. Wafer fabricators could either install costly new scrubbers or switch to a process requiring a smaller volume of chemicals.

Even in the best of circumstances, the threat of high-tech pollution will grow as production expands. New methods will bring new, unanticipated problems, as well as opportunities for improvement. The task of monitoring such hazards will require vigilance not only from public agencies and industry specialists but from the public and the press. The people of Silicon Valley, because of the tragedy of Los Paseos, appear to be aware of the problem, but massive public education is still necessary. Nationally, publicists and boosters have created a deceptively enticing image of the industry as clean and light. High-tech executives may be more willing than others to clean up their act once pollution is discovered; however, this only means that public pressure for both the prevention and clean-up of high-tech pollution *can* succeed. To generate that pressure, those outside high tech must keep tabs on the environmental impact of high-tech industry, or assuredly all of us will pay the costs.

Chapter Nine
THE ASIAN CHALLENGE

■ ■ ■ ■ ■

Prying open the cabinet of any microcomputer or video game reveals a collection of integrated circuits stuffed neatly into printed circuit boards. Each chip is marked with the logo of its manufacturer and stamped with the name of the country where it was assembled and packaged. The typical high-tech machine contains chips produced in the United States, Japan, Malaysia, the Philippines, and even El Salvador, as well as other components built in Taiwan, Hong Kong, and Singapore. The majority of the "Japan" chips were indeed made by Japanese-owned companies, but the rest are for the most part chips that were fabricated in the United States before being shipped abroad for assembly.

The companies which dominate high tech are truly multinational. Not only do many produce goods within major foreign markets; a large number also assemble overseas equipment and components for sale in the United States. The "American" semiconductor makers, for example, import more chips from their own Asian assembly plants than the United States receives from Japan. And they have roughly as many employees in Asia as they do in America.

The global dimension of "American" electronics production has so far either gone unseen or been ignored by most of the economists, politicians, and business leaders who tout high technology as the key to American reindustrialization. They urge tax, trade, educational, and economic policies to help American-based firms defeat "Japan, Inc." in the world marketplace, instead of promoting employment within the United States.

Although many industries operate on a global scale, no industry has so finely tuned its global operations as microelectronics. Integrated circuit production, in particular, takes place on a global assembly line. Manufacture consists of several distinct steps, each requiring its own mix of skills, capital, and materials. The shipping costs are slight, for chips and the silicon wafers from which they are made are valuable but small. Consequently, manufacturers can locate each step in nations or regions where production costs, in the long run, are likely to be lowest.

The production of integrated circuits begins in design laboratories, where engineers link up symbols representing microscopic transistors to form computerized patterns more complex than the map of Manhattan. From those patterns, a series of photographic templates, called masks, are formed. American-based companies conduct the bulk of their design in the United States, particularly in Silicon Valley.

Through photolithography or similar techniques, the mask patterns are projected onto thin silicon disks, called wafers, which range from three to six inches in diameter. Microscopic chemical "impurities" and metallic conductors are etched and baked into each wafer in super-clean rooms, where the workers wear surgical gowns and gloves to keep unwanted particles from spoiling a circuit. This process, called wafer fabrication, requires complex equipment and the support of specialists. All U.S.-based firms do their wafer fabrication in the United States and other developed countries.

The completed wafers are then shipped to East Asia, where they are diced into individual chips. With the aid of microscopes or more sophisticated, automated assembly machines, long lines of young women attach threads of gold wire, thinner than a human hair, to each chip. Then the chips are mounted and sealed into packages that resemble rectangular caterpillars. Assembly requires dexterity but very little training, so manufacturers have established plants or hired subcontractors in countries where the primary attraction is cheap, productive labor. The Malaysian government has even issued an invitation to electronics companies, which reads in part: "The manual dexterity of the oriental female is famous the world over. Her hands are small and she works fast with extreme care. Who, therefore, could be better qualified by nature and inheritance to contribute to the efficiency of a bench-assembly production line than the oriental girl?"

Masks, wafers, and chips are all carefully examined and tested, for errors serious enough to force the discarding of a chip are common. Most of this testing is carried out in the United States, although offshore locations with a well-trained but comparatively low-paid technical workforce, such as Singapore, are also becoming important test centers.

Offshore production is essentially as old as the integrated circuit. Fairchild Semiconductor set up its first foreign plant in Hong Kong in 1962; General Instruments established a semiconductor assembly facility on Taiwan in 1964; and Fairchild opened up South Korea in 1966. When the Mexican government's Border Industrialization Program geared up in 1967, Motorola established a factory there.

At first, semiconductor companies chose to set up shop in places where the American manufacturers of consumer electronics—radio, television, tape recorders, etc.—had already broken ground. Both industries moved offshore to cut labor costs, but there were two main differences. First, the home electronics producers had moved offshore to compete with low-priced Japanese goods, but Japan did not become a factor in the global chip marketplace until the mid-1970s. In other words, American chip makers moved abroad to compete against each other. Second, the consumer electronics industry spawned a number of indigenous Asian enterprises, but semiconductor firms have chosen in most cases to maintain 100 percent control over chip assembly, since assembly is only one step in a larger production process.

By the end of the 1960s, the chip-making industry was growing much faster than consumer electronics, and producers sought new offshore locations, including Singapore, Malaysia, Indonesia, Thailand, and the Philippines. Meanwhile, a few of the older assembly sites were shut down. Texas Instruments, for example, closed its Curaçao factory in the face of labor troubles.

Beginning in 1972, Asian capitalists, sometimes with the cooperation of Americans, established a handful of independent assembly and testing subcontracting facilities. In fact, the first semiconductor companies in the Philippines were locally run subcontractors. These firms handle overflow production for major merchant semiconductor firms, as well as assembly for new chip makers and chip users engaged in captive (in-house) production.

Manufacturers gradually learned the relative advantages of working in each country. Labor proved cheapest in populous, poor countries like the Philippines and Indonesia, but Hong Kong and Singapore offered better trained, more experienced managerial and technical workforces. Some firms, such as National Semiconductor and Fairchild, established a division of labor within Asia, with Singapore and Hong Kong facilities specializing in administration, sales, and testing. Thai, Indonesian, and Philippines factories, on the other hand, conduct labor-intensive assembly on large, standard production runs—what is known in the trade as "jellybean" work.

Today, there are about a quarter of a million non-Japanese Asians, primarily young women, working in the semiconductor industry. Many important chip makers have more workers abroad than in the United States. National Semiconductor, for example, reported in 1982 that 22,700 of its 38,300 employees were in Southeast Asia. Though semiconductor shipments from Japan to the United States are growing, they still represent only a fifth of the U.S. market. Most imported semiconductors come from Malaysia, the Philippines, Singapore, and South Korea.

U.S. trade laws have encouraged this international division of labor. Products which originate in the United States but are processed abroad are subject, upon reimportation into the United States, only to customs duties on that portion of product value added abroad. Thus, semiconductor manufacturers have paid tariffs on the value of their foreign assembly and testing rather than on the total product value. Since value added overseas has at times exceeded half the value of a typical chip, the savings were substantial. Of the $5 billion worth of semiconductors imported into the United States in 1983, more than two thirds qualified for special reimport treatment.

Starting in 1985, reimport tariff subsidies will no longer apply to semiconductors, but only because chip makers are getting an even bigger break: all duties on semiconductors, including value added abroad, are being eliminated. Semiconductor makers expect to save at least $100 million a year in duties.

Some observers have predicted that the gradual automation of semiconductor chip production will make domestic assembly competitive, as labor costs decline relative to other factors of production. Indeed, a few major chip producers, such as Intel, Fairchild, and Motorola, are starting up domestic assembly facilities.

Nevertheless, unless conditions change unexpectedly, the bulk of assembly work continues and is likely to continue overseas. Automated equipment still requires operators and technicians, and those workers come cheaply in the Far East. Consequently, over the last decade, manufacturers have gradually automated their Asian plants. Some companies may break in their new, highly complex automated systems at domestic plants, but once the processes are established, they will adopt those techniques abroad, too.

Chip makers have maintained a relatively consistent balance in their division of labor between the United States and the Far East, expanding both their domestic operations and foreign employment at the same time. Now employment is leveling off alike at home and abroad, because automation is reducing the need for workers at all stages of production nearly as fast as the market grows and output is increased.

By contrast, American computer manufacturers historically have done most of their production in the United States. Japan is the largest foreign source of data-processing equipment sold in the United States, but it supplies only a fraction of the market.

In the last few years, however, the production of computer components and peripherals, such as video display tubes, keyboards, microcomputer disk drives, and printers, has been moving overseas. Disk drive manufacturers like Tandon and Seagate are locating in Singapore, which has a strong precision machining industry as well as labor costs far below those in the United States. Manufacturers of video display monitors and terminals, on the other hand, are settling in South Korea and Taiwan. Both those countries, which have established successful television manufacturing industries, have the skills and suppliers to produce monitors and terminals cheaply.

The movement of computer and peripherals manufacturing to East Asia differs from the expansion of offshore semiconductor employment. Semiconductor makers have historically hired production workers in the United States at roughly the same pace as they grew abroad. But numerous equipment manufacturers—including Atari, ITT-Qume, Tandon, and Seagate—have recently laid off hundreds or thousands of domestic production workers, in some cases actually closing down plants in California.

The continuing role of Asian production workers in high-tech industry is an irony of major proportions. Many of the world's most sophisticated industrial products are assembled by the world's lowest-paid industrial workers. Wages in the Philippines, Thailand, and Malaysia average about U.S. 50¢ an hour; in Indonesia they're even lower. The women who assemble chips and other high-tech products live far from the fruits of their labor.

High-tech employers may treat their American professional and technical employees well, but the industrialists who claim to be bringing us the "second industrial revolution" treat their Asian workers in a style reminiscent of the first industrial revolution. Quite openly, high-tech employers in most countries hire only young, single women for assembly tasks, and they rely on quotas, threats, bonuses, and games to maintain a dehumanizing pace of production.

In some countries wages rest at the legal minimum, while elsewhere electronics pay is higher than work in other industrial sectors, such as textiles. In either case, electronics production does not pay enough for the average worker to support a family, which explains why firms seek out single women. Unions which represent the demands of workers are not tolerated.

Workers are faced with many of the chemical hazards found in U.S. plants, but the biggest hazard is microscope work, which usually involves the inspection of circuits or the bonding of gold wires to semiconductor chips. Microscopes, especially when used intensively all day six days a week, can cause dizziness, eye strain, and tension, as well as deteriorating vision. The women, who need 20/20 vision in order to get their jobs, frequently end up wearing glasses after only a few years. In Hong Kong, most assemblers over age twenty-five are called "Grandma" because they wear glasses.

In the mid-1970s, Dr. Son Jun-Kyung, head of the Ophthalmology Department at Paik Hospital in Seoul, South Korea, conducted examinations on sixty-four women employed by three American microelectronics plants. He found that 47 percent suffered from near-sightedness, conditions caused by long hours of concentrated eye-work on small objects, and 19 percent from astigmatism. In addition, 88 percent had chronic conjunctivitis—eye inflammation—apparently caused by the presence of toxic gases or dust in the factories.

The increasing use of semi-automatic bonding equipment, installed to raise productivity, has reduced the amount of scope work, presumably reducing the hazards to workers' eyesight. Yet the increased pace of automated production work may be a hazard in itself. Some firms pay extra to scope workers, but the women are sent back to other jobs as soon as their productivity or accuracy fails.

Instead of offering a living wage, employers attempt to boost morale and production by piping in Western music, selling cosmetics, giving away company T-shirts, and sponsoring beauty contests and volleyball leagues. Factory women, some of whom must share their dormitory beds with workers on other shifts, are blessed with the opportunity to be chosen "Miss Free Trade Zone" or "Miss Dual In-Line Package."

Despite the low pay and tedious work, people still flock to high-tech factory jobs. Air conditioning, installed for the good of the chips, is one attraction, but the main reason young Asian women seek electronics assembly employment is that there really isn't much work for them anywhere else. Outside of Singapore, most developing Asian countries have large labor surpluses, so there is little respectable wage labor for young women. In most cases, these workers need the money to supplement their meager family incomes, but some women hope that industrial employment will offer them independence from the traditional, patriarchal relationships. Often, sadly, the authoritarianism of the factory becomes a substitute for authoritarian village life.

Employers tout the dexterity of their young employees; however, another traditional characteristic of Asian women, docility, appears to be more important. Factory "girls" are expected to work hard, for long hours, for little money, without complaint. Yet conditions are frequently so bad that even they rebel, and even in countries in which dissent is not normally tolerated.

In 1982, for example, workers at Control Data's Korean subsidiary not only went on strike for better pay but "kidnapped" company negotiators from the United States—for several hours, they refused to let two Control Data officials leave the negotiations. Fairchild workers walked off the job in Indonesia in 1981. The two largest assembly subcontractors in the Philippines, Dynetics and Stanford Microsystems, were both hit by strikes in 1980. In 1978 and again in 1979, Fairchild's Hong Kong assemblers struck for—and got—higher pay. And in 1977, workers at Signetics in Korea staged a sit-in in the company cafeteria.

In Malaysia, workers have responded in an effective, though definitely less conscious way. Sometimes when production pressures are especially high, "mass hysteria" breaks out. First one woman is possessed by a "spirit." Soon the whole workforce is weeping and writhing in what is essentially a subconscious, socially accepted wildcat strike. Employers frequently bring in traditional healers, "witch doctors," to get production moving again.

The Malaysian workers may shortly have an alternative to hysteria, since in 1983 the government finally recognized the right of some electronics workers—not yet those doing semiconductor assembly, however—to join unions.

If the Malaysian government actually sanctions labor union rights, it will be the only East Asian country, other than Japan, that consistently tolerates genuine union activity. Hong Kong, still a British colony, tolerates unions as well, but the leading labor organizations there, which are closely aligned with Communist China, have shown relatively little interest in struggles over wages and working conditions at their electronics plants.

In other countries, protesting or striking workers and their supporters have been beaten, jailed, or deported by security forces. Though employers often use more sophisticated approaches to combat labor organizing, government repression underlies all industrial relations in most of Asia. Governments artificially hold down workers' wages through conscious policies of social control.

In fact, it appears that the widespread suppression of dissent in some countries, such as the Philippines and South Korea, is the direct consequence of economic policy. Those governments have adopted the strategy of export-led industrialization, which means that they have targeted their national resources to promote manufacturing industries, such as textiles and electronics, which export to the United States, Japan, and other developed countries. In order to attract foreign investment in those industries and to make their products competitive, they must keep wages down. Independent labor organizing and strikes, if permitted in these countries, would boost wages to the point where employers would look elsewhere for cheap labor, so dooming the export-oriented approach to development. Only if all countries were to permit wage-raising organizing activity could economic human rights become compatible with export-led industrial development.

For now, there is no way to abandon the policy of social control without abandoning export-led development. But that does not mean giving up on industrial growth. The most proven path to economic development, as well as the most equitable one, is to pay workers enough so that they can afford to buy what they make. This is how the United States, Japan, and other developed countries industrialized. Such an approach does not mean the abandonment of trade. But it does mean that countries following this path will export only those goods and services which they can produce competitively without using brute force to suppress wage rates.

Despite the phenomenal increases in gross exports, providing low-paid assembly work for high-tech corporations has not proven all that beneficial to most East Asian countries. In high tech, in particular, the foreign exchange earnings of nations hosting assembly operations are minor. Most of the parts and technology are imported—and much of the earnings are exported. In Thailand, for instance, a government study showed that 63 percent of the value of integrated circuits exported went directly for imported parts, wafers, and so on. Profits made up another 16 percent, with an undisclosed share of the earnings sent back abroad to the parent companies.

Although the various forms of high-tech assembly have provided jobs for at least 300,000 young Asian women (outside of Japan), the industry has barely caused a ripple in the vast reservoir of unemployed and underemployed; in fact, since most companies recruit young women from outside the main labor force, the effect upon unemployment is minuscule.

Furthermore, high-tech production has not transferred technology to Asia along with its assembly plants. Assembly requires little technology and no scientific training. It is unlikely that Indonesia, Thailand, the Philippines, or even Malaysia will ever have their own chip design or wafer fabrication industries.

South Korea, Taiwan, Hong Kong, and Singapore may, on the other hand, successfully establish their own, indigenous high-tech industries. But they have been forced to make an end run around the assemblers, buying the technology outright. Taiwan's Industrial Technology Research Institute, for example, hired RCA, which has substantial consumer electronics investments on the island, to provide the equipment and training for its wafer fabrication.

More recently, South Korea's largest consumer electronics manufacturers—Gold Star, Samsung, Hyundai, and Daewoo—have all decided to move into wafer fabrication. Though all four have either established or planned fabrication, as well as assembly, in South Korea, each initiated their chip-making ventures by setting up shop in Silicon Valley. By paying a relatively high price, the Korean companies are finally obtaining the technology to design and fabricate circuits similar to those which have been assembled in their own country for nearly two decades.

Without the educational and economic infrastructure of Japan or the United States, however, the Koreans will be unable to compete effectively in the world market for advanced circuits. They do not have the resources to match the huge sums being spent on research and development in the United States and Japan, though they may do well making the less sophisticated, mass-produced chips that go into televisions, digital watches, and other consumer products. There is a market for those in Korea: the export-oriented consumer electronics branches of the same Korean electronics firms.

The Koreans are not the only foreigners buying their way into Silicon Valley. Indeed, it is hard these days to define an American semiconductor firm. Philips, the Netherlands-based electronics giant, bought Signetics in 1975; French-owned Schlumberger, a technology-oriented oilfield services firm, took over Fairchild in 1979;

and Nippon Electric, the largest semiconductor maker outside the United States, bought Electronic Arrays in 1978.

■ In most cases, foreign firms—as well as a number of U.S.-based multinationals—have invested in Silicon Valley as their entry into the world of high tech. For Japanese corporations, however, operations in Silicon Valley are a small part of a much larger, systematic strategy to adapt and improve on the American technology. Japan's government and major electronics companies have, in fact, made high-tech parity or superiority over the "American" electronics industry a national priority.

The Japanese electronics firms first established themselves in the international marketplace by mass-producing transistor radios. Until the mid-1970s, the Japanese, like their South Korean counterparts in the 1980s, primarily exported consumer electronics goods. Even today, Japanese firms dominate the world market for video cassette systems. Unlike Korea, Japan itself is a major market, providing an economic base for investment in production for other markets.

The Japanese took advantage of that base during the recession of 1975–76, when U.S.-based merchant semiconductor firms scaled back their investments. Japanese companies, taking a longer view, expanded, and when the market for chips opened up again, they hit the ground with a running start. As the demand for computer memory chips—at that time, the 16,000 bit, or 16K RAM (random access memory) chips—skyrocketed, American manufacturers were unable to supply the market. Even National Semiconductor, which complained of rising Japanese competition, bought chips from Japanese firms. By 1980, Japanese companies had won 42 percent of the American market for 16K RAM chips.

As the Japanese established themselves in the U.S. market, American chip buyers learned that the quality of Japanese semiconductors exceeded that of the American versions. In 1980, Hewlett-Packard, primarily a manufacturer of instruments and computers, conducted reliability tests on circuits supplied by a number of companies. Much to the chagrin of its Silicon Valley neighbors, Hewlett-Packard announced that the Japanese quality was superior. Though American chip makers have since improved the reliability of their circuits, Hewlett-Packard reports that Japanese firms have also moved forward, so maintaining their lead.

Meanwhile the Japanese government, through the powerful Ministry of International Trade and Industry, targeted the chip industry for an all-out assault on the world market. In 1976, it launched a four-year research project into the design and manufacture of very large scale integrated (VLSI) circuits, bringing major Japanese electronics firms together into a technology-sharing research consortium. During that period, the Japanese government provided the industry with hundreds of millions of dollars in loans and grants for semiconductor development.

As producers moved into the next generation of memory chips, 64K RAM's, Japanese firms took an even larger share, 70 percent, of the initial American market. More important, in 1981, the Japanese launched another government-subsidized high-tech research program. Nicknamed the Fifth Generation project, the new ten-year, $400 million program was designed to help Japanese companies leapfrog to the leadership of the computer industry by generating breakthroughs in both hardware and software design.

So in 1977 U.S.-based semiconductor companies circled their wagons to form the Semiconductor Industry Association, ostensibly in response to the Japanese challenge. Charles Sporck, head of National Semiconductor and a member of the Association's board of directors, warned that unless the federal government took action, the domestic semiconductor industry would be "overrun and destroyed within ten years." Association spokesmen argued that America's economic strength depended upon its leadership in microelectronics technology. Jerry Sanders, founder of Advanced Micro Devices and another Association director, has called semiconductors the "crude oil of the 1980s."

American firms charge the Japanese with unfair competition. They decry the Ministry of International Trade and Industry's targeting efforts, and they complain that the Japanese market is closed to their products. They also blame a financing system which gives Japanese firms easy access to loans for expansion.

In general, the Japanese have competed fairly, although in some instances the rules that govern their behavior differ from the norm in America. The ministry's VLSI project did help Japanese manufacturers, but the U.S. semiconductor industry, particularly in its early days, has benefited much more from government procurement and research contracts. The VLSI project, while focused to solve important technical problems, represented only a fraction of the annual private sector budget for chip research in the United States. IBM, for example, spends a substantial portion of its $2 billion-plus annual research and development budget on semiconductor design and processing. Furthermore, although U.S.-based companies criticized the Japanese effort, several, such as Fairchild, sold production and test equipment directly to the project.

Throughout the late 1970s, Japanese tariffs on semiconductor imports exceeded those of the United States. When it became clear that the Japanese had become the equals of their American competitors, the Semiconductor Industry Association demanded that duties be equalized, and the Japanese agreed. In fact, in 1984 the two governments announced agreement in principle to eliminate duties on most semiconductor imports.

Breaking informal barriers against U.S. chip sales in Japan has been more difficult. Nippon Telephone and Telegraph, Japan's telecommunications monopoly, has agreed to purchase American equipment, but it, as well as other Japanese semiconductor users, still appears to favor Japanese suppliers. However, what American vendors interpret as national favoritism may simply be the Japanese custom of sticking with long-standing customers.

Furthermore, since the major users of chips in Japan are the largest producers, they "buy" components in-house. This puts the American semiconductor firms at a disadvantage, but it is an obstacle that faces Japanese as well as U.S.-based open-market firms in the United States. Many of America's large users of semiconductor products, such as IBM, AT&T, and General Motors, have traditionally supplied most of their own chips as well.

The Semiconductor Industry Association views with envy the large pool of low-cost loan capital available to Japanese firms. Big Japanese companies typically rely much more heavily upon debt to finance their operations than U.S. firms. In exchange for low interest rates, they accept more outside control and lower returns on stockholder equity than their American counterparts. The Association considers those financing methods a Japanese advantage, but no one in the U.S. high-tech industry is suggesting that the U.S. government and banks should copy the Japanese. Rather, high-tech spokesmen propose other policies designed to promote high-tech investment.

Despite this flood of rhetoric about the rise of the Japanese high-technology industry, the U.S. semiconductor industry is still the first in the world. Japanese semiconductor exports to the United States surpass U.S. shipments to Japan, but official U.S. statistics exaggerate the Japanese advantage because they ignore shipments to Japan from American subsidiaries in Asia.

While the Japanese have become the major suppliers of memory chips, the American producers still dominate the supply of more sophisticated chips. Intel alone, for example, held a remarkable 70 percent share of the Japanese microprocessor market in 1983. And memory chips, although they make up a big chunk of the global semiconductor market, are less profitable to produce and market than most other integrated circuits.

Moreover, Japanese companies lag far behind the United States in computer sales. Worldwide, IBM's sales dwarf the data-processing sales of all its Japanese competitors combined. Skilled as they may be in mass production and marketing, the Japanese high-tech enterprises have yet to prove themselves competitive in either hardware or software design. Most top Japanese computer manufacturers market "plug compatible" machines, designed to run standard IBM software.

In the computer business, in fact, it is still more accurate to divide the world of competition into IBM versus everyone else rather than the United States versus Japan. IBM is a global firm, based in the United States, of course, but it has been operating in Japan since before World War II. Until 1979, IBM was the largest Japanese manufacturer of computers; today, it remains close behind Fujitsu.

Nevertheless, Japan, through hard work and an industrial system that differs from that of the United States, is running a close second in the global high-tech economic race. Intel's Robert Noyce, co-inventor of the integrated circuit, suggests that Japan's growing industrial strength is a direct consequence of the dawn of the information age. Japan lacks the land and natural resources critical to competitiveness in the agricultural, extractive, and heavy industrial sectors. But its skilled labor force, supported by social investment in technology, gives Japan a comparative advantage in electronics.

To maintain American economic growth and employment, the United States must move ahead just as fast as the Japanese. To most high-tech executives, that means a federal industrial policy, but not the targeting approach of Japan's Ministry of International Trade and Industry. High-tech companies have proposed a complex web of government programs, ranging from ongoing tax incentives for research and development to strengthening training programs for engineers. Some of their less controversial proposals, such as the expansion of copyright law to cover chip designs and the clarification of anti-trust laws to permit cooperative research programs among high-tech competitors, have already been enacted.

■ Although American high-tech firms advocate policies designed to strengthen their hand against the Japanese challenge, they have not yet suggested ways of taking on the other Asian challenge, offshore manufacturing by U.S.-based firms. At this time, it is perfectly legal for high-tech firms to accept American tax subsidies for product development, and then turn around and assemble the new products in Asia.

The existing American industrial policy, which is supported by the U.S. taxpayer in the form of either government spending or tax breaks, does not presently include the generation of domestic production employment among its leading goals. Such a focus is necessary, however, for without an assist from public policy it is unlikely that high tech will be able to generate enough jobs to replace those of the machinists, telephone technicians, bookkeepers, and so on who are being displaced by microprocessors, computers, and other high-tech products.

Employers argue that overseas assembly has allowed them to lower production costs, boosting the market for their goods and consequently increasing domestic employment in work like wafer fabrication. However, the cost savings from cheap Asian labor have been minuscule when compared to the exponential decline in cost-per-circuit-element resulting from advances in design and production technology. It is that amazing decline, not offshore production, that has driven the growth of the semiconductor market.

Employers also claim that offshore production has made it possible for them to compete with foreign producers, particularly the Japanese. But the American semiconductor firms assembled their products abroad long before the Japanese were a serious factor in the market. Now that the Japanese are competitive, the claim has an element of truth. Yet all major Japanese chip manufacturers have learned how to compete effectively while assembling most of their chips at automated plants in Japan.

If the tax breaks offered to American firms were made contingent upon the establishment or expansion of domestic assembly facilities, high-tech production employment in the United States would jump. In that case, the cost of high-tech subsidies to the U.S. Treasury would be justified and perhaps paid back by an increase in American employment.

Furthermore, goods produced in foreign plants—owned by U.S.-based as well as foreign firms—should not be subsidized by practices that artificially suppress wage rates abroad. The most significant of such practices are "internal security" measures, which prevent unionization and free collective bargaining.

In most of Asia, anti-labor repression is dependent upon the diplomatic support or tolerance of the United States, and even more important, upon American military aid. Industrial pay in countries like the Philippines, Indonesia, and South Korea would probably rise soon after any reduction in U.S. backing for a dictatorial rule.

Or the United States could deny subsidies, such as insurance from the Overseas Private Investment Corporation, to American firms operating in countries that do not guarantee the fundamental rights of workers. In 1984, Congress linked the reduction of duties on imports from poor countries, under the Generalized System of Preferences program, to the observance of workers' rights. The Preferences do not apply to electronics products, but similar provisions could be added to the reimport subsidies exploited by semiconductor manufacturers for more than two decades. With the total elimination of semiconductor duties in 1985, a reimport subsidy rights standard would have no impact on chip production, but it could force countries involved in other kinds of electronics assembly to permit some form of genuine workers' movement. If they refused to change, the loss of subsidies would handicap their ability to export.

Sanctioned organizing would bring more strikes, higher pay, and better conditions. As overseas costs increased, some work would return to the United States.

Asian countries, faced with a higher quality but a lower quantity of jobs, would then reassess their dependence upon industrial exports. Thus the division of labor, on the global assembly line, would be recast.

The real Asian challenge to high-technology industry is a challenge to our economic system. Can high-technology industry succeed in simultaneously stimulating employment at home and economic development in Asia?

■ Meanwhile, the apparent economic conflict between Japan, Inc., and America, Inc., may devolve into competition among industrial partnerships that link U.S. and Japanese enterprises. Those partnerships, which are likely to increase over the next decade, include anything from technology-sharing agreements to direct investment. For example, in the semiconductor industry, LSI Logic (a small Silicon Valley firm) and Toshiba have jointly devised a specialized memory chip. Hitachi supplies Hewlett-Packard with memory chip production technology, and Fujitsu bought a license to produce Intel's most advanced microprocessors. Nippon Electric, through its wholly owned Electronic Arrays subsidiary, produces chips in the United States, while Texas Instruments operates three factories in Japan. Texas Instruments actually exports memory chips from Japan to the United States.

Trans-Pacific cooperation is widespread in the computer industry as well. IBM's Japanese subsidiary is essentially run by its Japanese staff, while Japanese computer firms export to the United States through American partners. Fujitsu, for instance, sells its machines through Silicon Valley's Amdahl Corporation, in which it holds a 49 percent interest. Nippon Electric supplies the American market through Honeywell, a fixture in the U.S. computer market for decades. Japanese microcomputer makers went to Microsoft—the firm that supplied the IBM PC with its initial operating system—for a standardized operating system. And Epson, a Japanese maker of components, computers, and printers, has set up a software subsidiary and a computer research lab and listening post in Silicon Valley.

National Semiconductor's situation clearly illustrates the contradiction between "trade war" rhetoric and the reality of U.S.-Japanese cooperation. Though National's top executives have repeatedly sounded the alarm about the Japanese threat, the company's future depends heavily upon its working alliance with a Japanese competitor. National relies upon sales of Hitachi-made computers for a large chunk of its profits. In this light, IBM represents more of a threat to National than the Japanese Ministry of International Trade and Industry does. Indeed, when Hitachi was caught with pilfered IBM computer designs in 1982, at least one IBM document turned up at National's computer division.

The fight to dominate high tech is not a battle between two nations but a complex race among inter-company partnerships, many of which link firms based in the United States and Japan. Americans must therefore look beyond the address of company headquarters when fashioning industrial policy. In promoting high-tech industry and employment, we must carefully examine both the players and their interests.

Chapter Ten TRADING SECRETS

■ ■ ■ ■ ■

In 1982, Peter Carey, an award-winning investigative reporter for the San Jose *Mercury News,* decided to check out personally the U.S. government's controls on the export of high technology. He flew to Singapore, taking with him a box which he had obtained from Intel Corporation that was marked up to look as if it contained advanced integrated circuits. No one took any notice of the box until he got to Malaysia, where customs officials confiscated its actual contents, a bunch of newspapers. As Carey found out, those who try to control the international flow of information have fallen far behind the means of transferring it.

Since the beginning of history, human beings have

recorded, stored, traded, and used information. Though important, it was not subject to frequent exchange. Its value was subjective, and like that of an oil painting, not easily quantified. Today, however, information has become a clearly defined commodity. The development and transmission of recorded knowledge, ideas, and data is central to the workings of our modern society. Its value is measurable and, in aggregate, enormous.

But one cannot possess information, in the usual sense. It has essentially unlimited value. Whereas a car can only be owned by one person or shared by a small group, millions of people can now use the same piece of information simultaneously. Coded electronically, or printed on paper, information can be copied and used at a fraction of the cost of its original production. Unlike reproductions of paintings, the copies have the same value as the original; indeed, copies of computerized data are generally as precise and replicable as the initial version. Consequently, information products, once developed, can be provided to large numbers of people at virtually no cost. In the words of Stewart Brand, "Information wants to be free."

This is true whether the information is in the form of a book, a video movie, or software for running an auto plant or designing a semiconductor chip. The tools of reproduction include paper copier machines, video recorders, pirate television antennae, and computer memory devices such as disk drives.

The "theft" of ideas, either by foreign spies or by domestic competitors, has never been easy to control, but the present explosion of information technology makes the piracy of both trade secrets and national security information extremely practicable. Today, it is possible for a single person to hold in his or her mind a technological concept which can influence the fate of a corporation or even a country. At the same time, with a computer it is possible cheaply, simply, and quickly to copy a complex computer program without leaving a hint that the reproduction has occurred.

■ High technology within the United States is an industry that has thrived upon the open flow of information. How could new companies devise remarkable new products without access to the ideas and experience of established firms? In Silicon Valley, the word "trade" in "trade secrets" is a verb.

Yet the cheap, unrestricted replication of ideas threatens the economic foundations of high-tech industry. The lines between research and theft are blurry. Who will invest in new products and new processes if competitors can develop a virtually identical device for a fraction of the cost, simply by copying or imitating it? Who will place new information products on the market when it is possible that a majority of the products' consumers will "steal" rather than buy them?

High tech's tradition of openness dates back to the earliest days of microelectronics. When Bell Laboratories demonstrated the first transistor in 1947, neither the military nor private industry attempted to clamp controls on the newborn semiconductor technology. Bell licensed the technology widely, stimulating further developments in design and production technology.

The military soon became the largest customer for transistors, yet it did not attempt to classify semiconductor technology at a time when the U.S. government was tightly restricting access to another new technology, nuclear reactors. The Pentagon did not fully comprehend the significance of the transistor until the basic technology was already in the public domain. Of course, if it had tried to classify the transistor as secret, it is likely that AT&T, Bell Lab's parent and at the time the world's largest private corporation, would have fought hard to retain access to its own invention for commercial use.

In fact, Bell Labs encouraged the widest possible dispersal of transistor technology. It publicized the invention, licensed the technology, and staged seminars. Brookings scholar John Tilton links the spread of transistor technology to anti-trust proceedings brought by the U.S. Justice Department against AT&T in 1949. Justice wanted to separate Bell Laboratories from AT&T's operating companies, and it eventually won a liberal licensing requirement in the 1956 consent decree that settled the case.

Texas Instruments (TI) and Fairchild Semiconductor, the two companies that dominated the next wave of semiconductor innovation, followed Bell's lead and liberally licensed their technology, as well. TI developed the first silicon transistor in 1954, while Fairchild pioneered advances in processing technology. Teams at both companies invented integrated circuits—chips containing multiple transistors—at the end of the decade. Anyone who wanted to produce semiconductors had to obtain licenses, for a price, from AT&T, TI, and Fairchild.

The widespread licensing of fundamental semiconductor technology stimulated the industry's rapid growth and innovation; but a less formal transfer of technology, the defection of key professionals from pioneering firms, proved even more important. In a rapidly changing field, human know-how is the core of innovation. Though university training is valuable, in the 1950s and 1960s no college taught what the early semiconductor researchers learned on the job.

TI's semiconductor lab was headed by Bell Labs alumnus Gordon Teal. William Shockley, who later received the Nobel Prize for his role in the invention of the transistor at Bell, launched Silicon Valley's semiconductor industry when he brought together a collection of dedicated young researchers at his commercially unsuccessful Shockley Semiconductor. Eight of those engineers and scientists soon left Shockley to form Fairchild, which in turn spun off National Semiconductor, Advanced Micro Devices, Signetics, and numerous other firms. Two of Fairchild's eight founders, Robert Noyce and Gordon Moore, started Intel a decade later.

The personal computer industry, itself an outgrowth of the semiconductor industry, later flourished upon the open exchange of technology. During the mid-1970s, industry pioneers traded ideas at Palo Alto's Homebrew Computer Club before designing proprietary machines.

Apple Computer, which dominated the educational and professional market for personal computers until IBM joined the competition, deliberately designed its Apple II computer with an "open architecture." That is, rather than conceal the machine's internal hardware and software design, Apple publicized it; it also invited individuals and independent firms to write programs and design peripheral equipment for the computer. Because Steve Wozniak, the Apple II designer, had a deep instinct for sharing his work with others, Apple stimulated the creation of a huge software library which increased the practicality of personal computing and rapidly established microcomputer production as a profitable business.

IBM reacted to Apple's success by introducing its own personal computer in 1982. The open-architecture IBM PC, backed by the world's dominant computer manufacturer, almost immediately became an industry standard. Independent programmers placed thousands of programs on the market within a year of the PC's appearance. Soon most existing and virtually every new supplier of personal computers was offering "IBM-compatible" machines—computers that could run software written for the IBM PC. The appearance of such a large volume of software, designed to run on a standard type of machine, in turn opened up a much larger market for personal computers, particularly in the business community.

Yet open technology has its limits. Semiconductor producers, for example, have always concealed certain types of proprietary technology from their competitors. Typically, a firm publishes or licenses general technological advances, but it makes available specific product designs to just a handful of producers, called "second sources." (By lining up a large, reliable second source, a manufacturer reassures potential buyers, who must design equipment to use the particular component, that there will be a steady supply even if the initial producer is having problems.)

The line between design and technology is as fine as the circuits on a chip; unlike those circuits, however, it is subject to interpretation. The history of the semiconductor industry is rife with disputes over the theft of designs and production techniques, usually occurring after one or more researchers at a semiconductor firm left to start a competing company.

In fact, Fairchild sued its first spin-off, all the way back in 1959. The founders of Rheem Semiconductor, who included Fairchild's first general manager as well as ten other key Fairchild employees, had taken Fairchild's "cook book," or process manual, with them. The two firms eventually settled out of court, but the cost of fighting the suit crippled Rheem.

The same story has recurred many times since in Silicon Valley, with only the players changing. In 1983, there were at least twenty cases under way in which firms were suing former employees for stealing trade secrets. Such litigation is so common that several lawyers and firms in the Valley actually specialize in trade secrets cases.

It has always been illegal to walk off with trade secrets, yet until Congress extended copyright protection to the photographic templates (masks) used in chip fabrication in late 1984, the piracy of chip designs was not only considered legal by many in the industry but was a common practice. In 1979, Intel president Andrew Grove showed a congressional subcommittee photographic enlargements of Japanese and Soviet memory chips that appeared identical to Intel products. L. J. Sevin, president of Texas-based Mostek, displayed a copy of a pirate version of a Mostek memory circuit. The copy contained certain useless microscopic features found on the Mostek original. Like a cartographer's deliberate errors, known as "zero population cities," the presence of those features confirmed the piracy.

Supporters of the legislation pointed out that for less than $50,000 a pirate semiconductor producer could steal a design which had taken the chip's originating firm years to develop, consumed thousands of workhours, and cost millions of dollars. Without having to support the cost of research and development, the pirate firm could undersell the innovating company. In the long run, the industry's most innovative firms could have been driven out of the business.

Computer makers, like chip manufacturers, also jealously guard their designs. Apple Computer is engaged in an ongoing, worldwide legal battle against cheap, unauthorized imitations, most of which appear to come from Taiwan. IBM nearly drove Eagle Computer out of business when it forced Eagle to stop selling computers using a portion of the IBM PC's Basic Input-Output System (BIOS).

Several of IBM's largest competitors in the market for large computers are entirely dependent on their ability to make their systems compatible with IBM's—that is, capable of using software designed for IBM machines. Each time IBM introduces a new generation of large mainframe computers, these manufacturers of "plug compatible mainframes" rush to reverse engineer IBM's hardware and decipher its software. This process is not cheap, and it takes time—time that IBM uses to build up its market share.

In the summer of 1981, Hitachi, a major Japanese manufacturer of plug-compatible mainframes, thought it could get a headstart in the race to duplicate IBM's yet-to-be-released 308X series of computers. Through a former IBM employee who had moved to the computer division of National Semiconductor, Hitachi's American marketing agent, Hitachi clandestinely obtained ten of the twenty-seven volumes which made up the "Adirondack" workbooks, the 308X plans. Several months later, Hitachi agreed to pay over half a million dollars for a long list of IBM documents and components. *Fortune* reports that when Hitachi senior engineer Kenji Hayashi entered a room containing several cartons of what he thought to be IBM secrets, he triumphantly ripped a sticky IBM-logo label from one box and stuck it on the front of his notebook. His moment of success was exactly that long. Two FBI agents stepped forward, told him that the episode was all over, and placed him under arrest.

The arrest of Hitachi employees, as well as several from Mitsubishi on similar charges, was an international scandal of major proportions. But in Silicon Valley, where most of the encounters took place, industry veterans took the disclosures in stride. The Japanese, apparently, had erred only by going a little too far, and by getting caught.

In an industry where rapidly matching a competitor's

product, and in many cases designing a compatible device, can mean the life or death of an enterprise, the theft of trade secrets will undoubtedly continue. Technology heists frequently leave little or no evidence that a crime has occurred, since with a copier or computer equipment, one can steal the information without removing the original. Or one can copy the technology by dissecting a computer or chip legally purchased on the open market.

Furthermore, law enforcement personnel are unfamiliar with the signs of high-tech theft. While a cop on the beat may be able to recognize a "hot" car stereo, he would be wasting the earning power of his education if he could identify a stolen chip, a counterfeit computer, illegally duplicated software, or stolen designs. Even the best prepared investigators cannot keep up with all the developments in high tech.

■ If it is difficult to control the unauthorized use of designs and software by high-tech companies, it is virtually impossible to stop the unauthorized copying of mass-market computer software written for personal computers. Piracy by computer users appears as common, and as acceptable, as drinking under Prohibition. People who would never steal a pencil from a stationery store do not hesitate to duplicate, for their own personal use, a software package worth hundreds of dollars.

Any microcomputer program can be copied from one magnetic disk to another in a few seconds merely for the cost of the disk, usually just a few dollars. As commercial software becomes more expensive, the savings and the consequent motivation for unauthorized copying increase. Unlike video- or audio tapes, which require two machines for making copies, all one needs is a single microcomputer. Furthermore, unlike photocopies or video reproductions, the quality of the copy is as precise as that of the original.

To find stolen programs, an investigator would need to gain access to the private computer libraries of average computer users. Even if he or she were to identify the software, there is no easy way to determine whether it is a "back-up" copy of a legally purchased program. And should the investigator determine that the copy is unauthorized, there would still be no easy way to identify the thief or middleman.

Personal computer software vendors rely upon the honor system, threats published on software packages, and numerous technical tricks designed to prevent copying. Few users even notice the copyright warning, and most, if not all, technical methods of copy protection can easily bc dcfeated. There are several inexpensive "back-up" programs on the market designed to overcome the latest in anti-theft schemes. Most advertise that they can generate legal back-up copies, yet there is no way to control their use.

The biggest obstacle to mass reproduction is distribution. There is relatively little mass marketing of stolen software, since large-scale distributors must advertise their wares. Nevertheless, computer users whose only goal is exchanging information, with no thought of profit, have built complex networks of unauthorized transfer. Some of these networks are made up of hobbyists; others simply bring together schoolteachers who traditionally have shared their meager classroom resources.

Through computer clubs and electronic bulletin boards, hobbyists transmit a vast amount of information, including a large volume of software that is in the public domain—that is, legal to duplicate. There is no easy way for investigators to determine which programs are "hot," or where pirated programs originate. It would take police-state tactics and the allocation of significant technical and personnel resources to law enforcement agencies to control such distribution channels.

Rather than fight piracy, some programmers actually exploit the tendency of many computer users to copy and circulate useful software. These small entrepreneurs, such as Andrew Fluegelman of Freeware, request voluntary payments if the user feels that the software has value. Fluegelman will send PC-Talk, a program which makes it possible to use the IBM PC to communicate with other machines (and transfer software between two computers!), to anyone who sends him a blank diskette upon which to copy it.

When the user loads PC-Talk onto his computer screen, he sees a message that reads: "If you have used this program and found it of value, your contribution ($25 suggested) will be appreciated. . . . Regardless of whether you make a contribution, you are encouraged to copy and share this program." Fluegelman explains that "electronic information is inherently easy to copy, and copy-protection schemes are all going to great efforts to go against a natural tendency."

Fluegelman and other practitioners of the Freeware concept probably suffer a higher rate of "piracy" than the major software firms. He estimates that about one in ten users sends money. But his distribution and legal costs are low, and the income appears sufficient to support his small operation.

In Australia, in late 1983, the free distribution of software went a step further. Apple Computer had brought suit in a lower Australian court to prevent the distribution of a Taiwan-made computer containing internal software pirated from Apple. The court ruled against Apple, finding that the codes could not be copyrighted.

For several months this one decision effectively eliminated all legal protections against the copying of software in Australia. Hobbyists rushed to take advantage of the void, distributing, at duplication cost, vast numbers of copies of standard microcomputer applications software packages, including the most popular commercial programs for word-processing and spreadsheet generation. Software Liberation, an organization of Australian hobbyists which advocates distribution of software at cost, argued that copyright laws are an artificial restriction on trade, designed to prevent price competition. They urged government funding of software development as an alternative to the commercial system. However, in May 1984 the High Court of Australia reversed the lower court decision, putting an end to the legal piracy experiment.

The public financing of microcomputer software development for free distribution to consumers is politically unlikely anywhere in the advanced capitalist world, but the present open market is itself built on weak foundations. If software producers cannot recoup their expenses and salaries, then they will not be able to write new programs.

In the short run, the personal computer software market will probably split into two tiers. At the upper end will be programs, in most cases designed for business users, which will cost hundreds if not thousands of dollars. The companies which sell those programs will jealously guard their secrets, both by extensive legal action and by employing constantly changing anti-duplication techniques. They will never stop piracy, but they will scare enough users into buying to make a solid profit from their work.

At the lower end will be the low-priced programs, ranging anywhere from $25 to $100. Most of this software will be offered by firms and individuals who use normal retail channels; but Freeware-type marketing schemes fit in here as well. The major protection against the illegal copying of lower-tier software will be the low price of the product. Few users would justify or feel the need to steal programs in that price range. If the low end of the market becomes dominant, then piracy will have fulfilled a free-market function. It will have driven down the price of software.

In the long run, as the production and distribution of information commodities such as software make up an increasing share of American economic activity, the economic system will be put to a new test: How can private enterprise earn profits on commodities that "want to be free"?

■ The U.S. government, like the software firms, is in the business of trying to control the distribution of technology; but it is aiming to keep advanced know-how, including computer and microelectronics technology, out of the hands of the Soviet Union at any price. So far controls have proven ineffective, since information knows no national boundaries. In fact, the federal government's crackdown on the transfer of high technology has done much to injure the Pentagon's long-standing working relationship with both high-tech industry and America's leading universities.

There is no doubt that the Soviet Union and its allies have mounted a massive program—overt and covert, legal as well as illegal—to obtain electronics technology from the West and Japan. The Soviets seek computers and components not only for military missions but for education, industrial use, air traffic control, and virtually every application that capitalist nations have found for high technology, except home computing.

The USSR has long demonstrated an ability to mass-produce most military commodities, but it lags years behind the United States in the design, manufacture, and use of integrated circuits and computers. For example, the Soviets now make a microcomputer called the *Agat.* A bright red machine, it is compatible with the aging Apple II but 30 percent slower; it sells for the equivalent of $17,000, about twenty times as much as current versions of the Apple.

Presumably, the Soviets would import vast numbers of advanced computers and sophisticated chips if the United States, Japan, and their allies would sell them, but volume shipments are easy to detect and stop. Consequently, the Soviets have placed a heavy emphasis upon the collection of technological information and production equipment that would allow them to build their own.

Meanwhile, American manufacturers have sought policies which would satisfy government officials dedicated to handicapping the Soviets while still allowing the export of their products to the Eastern bloc. A Defense Science Board task force, headed by J. Fred Bucy of Texas Instruments, recommended a compromise in 1976. It suggested that the Pentagon oppose the export of "critical technologies"—that is, how to do things—rather than focusing on end products.

Despite widespread support for the Bucy Report in both the federal government and the electronics industry, the critical technologies approach has had little practical impact. The Department of Defense did not issue a public version of its Militarily Critical Technologies List—a 700-page, two-thirds secret reference guide—until late 1984, and it still restricts the export of many integrated circuits and computers.

To prevent the export of high-tech components and equipment, the government operates a complex licensing system, in which several agencies must review the export of electronic goods. Requests to ship large and medium-sized computers, complex telecommunications switches, state-of-the-art integrated circuits, and semiconductor production equipment are denied. Even shipments to U.S. allies must be licensed, although those are routinely approved.

Established high-tech companies usually cooperate, but over the past decade there have been numerous attempts by small companies, middlemen, and front operations to smuggle restricted equipment to the Soviet bloc. For example, in 1976 the FBI charged executives of Illumination Industries, a Silicon Valley producer of semiconductor production equipment, with attempting to smuggle $3 million in automated processing equipment to the Soviet Union in crates marked as washing machines and ovens.

As Peter Carey of the San Jose *Mercury News* confirmed, small items which are available to the American consumer are easy for foreign agents to obtain and smuggle. Microprocessors, for example, can be purchased at thousands of electronics hobby shops such as Radio Shack, and it is virtually impossible to halt the export of the tiny chips, which are so small that they can fit undetectably into a coin purse.

It is also difficult to prevent the Soviets from obtaining small numbers of advanced personal computers, yet the United States still regulates personal travel with microcomputers. One needs a validated export license in order to take a computer or software worth more than $1,000 out of the country. The United States has relaxed controls on some machines, implementing a 1984 accord with its allies; but certain common machines, such as Apple's Macintosh and IBM's PC-AT, still may not be shipped directly or indirectly to the Soviet Union or its allies.

To stifle the illegal flow of large high-tech goods, the

Customs Service has teamed up with law enforcement agencies to establish Operation Exodus, the sole purpose of which is to investigate possible violations of export control laws and regulations. Exodus relies upon conventional law enforcement techniques, but federal officials are planning a new investigative venture, called Project Rampart, the details of which are still classified secret.

For Rampart, the Customs Service has obtained sensors to detect electronically the export of embargoed high-tech equipment. Cooperating manufacturers are building "bugs" into export-restricted equipment. Like the anti-theft devices used at libraries and department stores, Customs Service detectors will locate those bugs when crates of embargoed machines move through airports and other shipping centers.

Due to both enforcement and the cooperation of U.S. producers, it appears that few embargoed high-tech goods make their way directly from the United States to the Soviet Union. Instead, illegal exports usually follow a circuitous path through allied and neutral countries, including South Africa, West Germany, Austria, Sweden, and Singapore.

To prevent its allies from sending weapons and weapons technologies to Communist countries, the United States spearheaded the formation of COCOM (for Coordinating Committee) in the late 1940s. COCOM, which includes Japan and all NATO members except Iceland and Spain, develops common lists of commodities that are to be denied to the Socialist bloc. To avoid the scrutiny and control of the various congresses and parliaments, COCOM has always operated on a nontreaty basis. Working under a "gentleman's agreement," it has no enforcement powers, and more than once individual nations have forced revisions and exceptions to the lists merely by threatening to ignore them.

From the point of view of conservatives in the United States, COCOM is generally ineffective, since most American allies favor much less restrictive controls than the United States. Still, the United States does negotiate with its allies to develop common standards, since COCOM is the only game in town.

To prevent firms operating in COCOM nations from relaying embargoed high-tech machinery to the Soviet bloc, the U.S. has intensified its monitoring of high-tech exports to those countries. High-tech firms must obtain licenses, even to export to COCOM destinations.

Since nearly all applications to export to allies are approved, the industry considers the paperwork and delays unnecessary; indeed, industry spokesmen argue that the procedures benefit their foreign competitors, who are not subject to the same delays. In 1984, the electronics industry got the House of Representatives to ease such requirements in its version of the Export Administration Act, but the bill died when the House and Senate could not work out differences.

Although the licensing of exports to allied countries consumes a great deal of bureaucratic energy, with few denials, the recording of shipments does give investigators a tool with which to monitor illegal diversion. For instance, in 1984 officials examined embargoed equipment that had been seized in Sweden and West Germany. With the aid of licensing records—as *Aviation Week* reports—they found that at least three well-known American companies had sold restricted equipment to a blacklisted intermediary.

To control the flow of technology through neutral Western countries that do not belong to COCOM, the United States is threatening to cut technology transfers to those countries. For example, it may withhold export licenses for shipments to Austria, reportedly considered a "sieve" leaking high technology to the Soviet bloc, unless the Austrian government enacts new, enforceable controls on high-tech reexports.

The shipment of high-tech goods to the Soviet bloc from third countries is so widespread that U.S. controls may never be fully effective unless allies agree to adopt America's hard line. Specific shipments, particularly those which originate in the United States, may be intercepted, but the Eastern bloc will still be able to obtain, over time, the most important items on its high-tech shopping lists. Frequent negotiation of COCOM restrictions could bring about more uniform treatment, but in recent years the United States has repeatedly imposed more severe export limitations on its own.

In many cases, the problem is more fundamental than getting allies to block indirect shipments from the United States. Today, U.S.-based firms no longer hold a monopoly over advanced technology and products. When the U.S. government halts a shipment that does not clearly fall within COCOM's restrictions, allied producers are perfectly happy to make the sale.

In 1974, for instance, U.S. officials denied Fairchild permission to sell Poland the technology and equipment to build a plant for manufacturing simple calculator chips. Some time later, Fairchild officials reported that French sources had supplied the necessary equipment. In 1980, Japanese, French, and other western European producers rushed to offer substitutes for American computer exports to the USSR that had been blocked by the Carter administration.

Most U.S. government attempts to control the eastward flow of technology are not only ineffective but undermine the development of high tech within the United States. Spokesmen for manufacturers, such as Silicon Valley Congressman Ed Zschau, suggest that the restrictions still focus too much on actual products rather than on the technology. "Most of the militarily significant U.S. technology secrets that the Soviets get from the West," he contends, "are obtained illegally—by espionage, stealing, smuggling, or from open literature." The industry maintains that enforcement resources could be more effectively applied if fewer items were controlled.

It is indeed futile to invest law enforcement and customs resources in programs to keep specific high-tech products completely out of the Soviets' hands. The Soviets need obtain only a few copies of an advanced chip or a single computer to copy its design. And once they have the design information, they can replicate almost anything, although production may be costly and unreliable.

Since 1982, the Pentagon and the Commerce Department have been aiming their sights at the export of high-tech information as well as goods. The Commerce Department is drafting new regulations requiring the licensing of the export of some types of educational and scientific data. The term "exports," in this case, includes both the presentation of scientific papers at conferences attended by foreigners and the hiring of non-U.S. researchers.

But it is virtually impossible to control the flow of computer-coded technical information that can be instantaneously transmitted across national boundaries. There is, in fact, a direct link between computer networks used by the Pentagon, as well as by the most advanced centers of computer science research in this country, and the Soviet Union, via a little known center for mathematics research in Austria. From their home bases, Soviet scientists, like teen-age hackers, need only obtain passwords in order to gain access to advanced but unclassified American research.

To those who consider global scientific exchange a prerequisite for world economic and social progress, such international networks underscore the promise of high technology. But to the dominant faction in Washington, they pose a threat to American security.

U.S. agencies are attempting to restrict the direct export of high-tech information, though they realize that lines of a map do not deter the transmission of data. The Department of Defense believes that the only way to keep high-tech data and ideas out of the hands of the Soviets is to keep the information out of the public domain. It has therefore instituted the prepublication censorship of unclassified research reports on sensitive technologies. Pentagon security officials have censored and actually denied the presentation of unclassified technical papers prepared by university, industry, and military scientists for delivery at numerous high-tech scientific conferences. The Department of Defense now divides all new, "unclassified" technical documents into one of seven distribution categories. Those labeled "A" are for unlimited dissemination, while X-rated manuals, specifications, blueprints, software, and so on may be distributed only to those specifically cleared to receive export-controlled technical data.

Representatives of technical societies and leading research universities, such as Stanford, MIT, and Cal Tech, have reacted strongly against all attempts to restrict publication. Research controls threaten the tradition of open exploration valued by many institutions and researchers. In particular, a number of universities told the Pentagon that they would not accept research contracts requiring prepublication censorship. This was not an empty threat, since many of those institutions already refuse work requiring a security classification.

In 1982, the National Academy of Sciences issued a report that attempted to resolve the controversy. The panel, headed by Cornell University's president emeritus Dale Corson, received top secret briefings. Unlike industry spokesmen, the panel found that "universities and open scientific communication have been the sources of very little of this technology transfer problem." The panel suggested that research which policymakers want kept out of the hands of the Soviets should be clearly marked "classified," not categorized as unclassified but controlled.

The fight for control over the flow of scientific information is likely to continue indefinitely. Although in July 1984 the Defense Department withdrew its proposal to require prepublication review of manuscripts covering sensitive studies categorized as "basic research" in the Pentagon budget, it moved ahead to establish its export-control labeling system.

The Pentagon has also proposed restrictions on foreign graduate student participation in military-sponsored campus research, since a student's knowledge can easily be taken home—exported—in his mind or in his personal papers. University spokesmen argue, however, that such controls could devastate their graduate programs. In 1983, 56 percent of the doctorates in engineering issued by U.S. universities and 38 percent of the mathematical sciences doctorates were awarded to foreign nationals.

Just how well do the Soviets exploit Western electronics technology? To bolster the proposals to tighten restrictions, Pentagon officials have played up the significance of technology transfer. Information security specialist Arthur Van Cook told a U.S. Senate subcommittee: "The Soviets view the U.S. and several other Western countries as a continuing source of important and openly available scientific and technical information. . . . In some cases, their acquisitions satisfy deficiencies in Soviet technology such as smart weapons, electro-optical, and signal and information processing technology for Soviet air defense systems."

Richard DeLauer, Under Secretary of Defense, claims that the Soviet Union is able "to take U.S. technology and insert it into new weapons systems much faster than the U.S. military services can deploy the same technology." However, the Soviet military industrial complex is even more rigid and bureaucratic than its American counterpart. The Soviets may produce some weapons more efficiently than the United States, but that appears to be because they generally field simpler systems, less reliant upon high technology. There is no hard evidence to support DeLauer's assertion.

Frequently, U.S. spokesmen clearly exaggerate the direct military utility of exported high-tech equipment. For instance, when Swedish officials impounded a Digital Equipment "VAX 11–782" minicomputer, which is a sophisticated but general purpose machine, an American customs official alleged that the VAX could be used "for missile guidance or something like that." American newspapers dutifully repeated that charge in their headlines, though it is highly unlikely that the VAX could or would be used for missile guidance.

Other U.S. officials suggest that the Soviets are collecting Western scientific information and products to develop the technological base to build future weapons systems. For example, a declassified 1982 CIA study reportedly claimed that Western microelectronics technology "has permitted the Soviets to systematically build a modern microelectronics industry which will be the critical basis for enhancing the sophistication of future Soviet military systems for decades."

In reality, the Soviets want computer technology not only to build advanced weapons but to modernize their entire economy and educational system. Dr. Lara Baker of the Los Alamos National Laboratories told a U.S.

Senate committee that the Soviets lack the equipment to train young scientists, programmers, and engineers: "Many [Soviet] export licenses requests . . . are for computer systems going to universities or scientific research institutes."

Official warnings about the Soviet utilization of American technology miss the point. For years the Soviets have been collecting a great deal of American and other Western computer and chip know-how, yet they still remain far behind. Their civilian systems are undeniably obsolete by U.S. standards. There is no reason to believe that newly obtained technology will help the Soviets to leapfrog to the front of the high-tech pack.

Even high-priority Soviet military electronics technology lags far behind the United States. The Soviet air defense system, as demonstrated in two Korean airliner incidents, is ineffective; Soviet radar and countermeasures equipment has performed poorly in the Middle East; captured Soviet weapons lack the sophistication of American arms. The Soviets are either unable to exploit the Western technology they have obtained, or they have chosen not to.

The countries of the Eastern bloc cannot match Western computer technology for two key reasons. First, they have no mass market for chips and computers. This not only keeps their high-tech industry small; it also minimizes the number of young people exposed to computer hardware and software. Second, without a tradition of open information exchange, innovation and research are stifled.

If American and Japanese universities and high-tech companies were standing still, the Soviets could eventually catch up by obtaining and adapting imported high-tech knowhow. But the state-of-the-art is a moving target, propelled by fierce inter-company and international competition and the constant movement of innovators and ideas. No matter how fast the Soviets glean American journals and buy Western hardware, they will remain second or even third class.

Until we find a way to end the arms race with the Soviet Union, the United States can best keep advanced weaponry out of Soviet hands by tightly controlling the export of such weapons systems and by keeping its designs confidential. Proposals which go further are self-defeating and indeed appear to arise from ulterior motives.

There are many American policymakers, in both major parties, who view export controls as a way to punish the Soviet Union for its undesirable domestic and foreign policies. In 1974, for instance, the Jackson–Vanik Amendment, proposed by hawkish Democrat Henry Jackson, linked U.S.–USSR trade to Soviet emigration policy. Later, following the invasion of Afghanistan, the Carter administration cut off both grain and high-tech exports to the Soviet. Such controls may have hurt the Soviet economy, as well as sectors of the U.S. economy, but they did not influence Soviet policy. During the mid-1970s, both Soviet emigration and U.S.–USSR trade declined. The Soviets are still in Afghanistan, but U.S. farmers and manufacturers lost important customers.

The Reagan administration has adapted a new variation of the long-discredited right-wing program of economic warfare against the Eastern bloc. W. Allen Wallis, Under Secretary of State for Economic Affairs, told Congress: "We want to avoid economic exchanges, particularly transfers of technology, that contribute to the military potential of the U.S.S.R. and its allies or that *subsidize the heavily militarized Soviet economy and thus alleviate their difficult resource allocation decisions.* Our policy is not one of economic warfare against the Soviets. We do not seek the 'collapse' of the Warsaw Pact economies; we could not cause this in any case" (emphasis added).

High-tech industry, of course, does not want to bear the burden of economic conflict with the Soviets. *Aviation Week,* a leading voice of the military industrial complex, editorialized: "Tightened U.S. government controls over exports, technology transfer and technical information are turning an ideological crusade into an administrative fiasco." The magazine maintained that the campaign has offended U.S. allies and other friendly nations, triggered internecine warfare in the federal government, alarmed industry, aroused the ire of the academic community, and confounded technical societies.

Even Edward Teller, "father of the H-bomb" and a conservative supporter of President Reagan's arms program, is against restrictions on the flow of scientific information. He points out: "In the last third of the century the United States has lost its position in all military fields, most specifically in those where we practice secrecy. But in one field—which is not military but which has military application—we are ahead of the world and way ahead of the Russians. This field is electronics, especially computers, and it is in this field that we do not practice secrecy."

■ America remains the world's leading technological power, but there is no longer such a thing as "American technology." New technologies—whether generated in the United States, Japan, or elsewhere—are quickly disseminated around the globe. Just as the United States cannot confine chip designs within its territory, other countries cannot keep high technology out.

Information technology now transcends national sovereignty, yet it is unlikely to bring either world peace or global equality. High technology is still controlled by the powerful institutions which have the resources to develop, adapt, and exploit it.

Throughout history, governments, churches, and other large institutions have periodically attempted to control the dissemination of ideas. At times they have succeeded. Today, ideas are as crucial to economic growth and military power as they are to politics, culture, and religion, and there are forces that seek to control those ideas as well. But officials cannot totally control the flow. The same technologies that contribute to police and military strength, as well as to corporate economic power, are making it increasingly difficult for centralized institutions to halt or even channel the flow of information.

Government should encourage rather than limit the flow. In most situations, personal freedom, economic growth, and even national security will be served by policies which take advantage of the "natural tendency" of information to be copied and distributed.

Chapter Eleven
SEEDS OF LIBERATION

■ ■ ■ ■ ■

The miraculous silicon chip, like the Sirens of Greek mythology, sings an alluring song. It transforms industrial wastelands into beehives of success. Some Americans, upon rising in the morning, need no longer send the dog out after the morning paper. The computer terminal, through the night, has automatically selected its master's news stories and collected his "mail." In the classroom, high tech can motivate even the most recalcitrant student, giving him individualized, spellbinding attention. With a computer system costing less than half the price of an automobile, a manager can instantly calculate his past, present, and future inventory, sales, expenses, and profit.

However, if we—like the ancient mariners—seek solace or liberation in the promise of high technology, then we too may ultimately perish on the rocks. Despite the universal appeal of computers and microelectronics, high tech is fragmenting society. To the wealthy and highly educated, high tech is a source of power, status, and comfort. To the dispossessed, it presents one more intimidating obstacle to advancement.

Computers and chip-based communications devices are proliferating, binding Americans ever more tightly into a vast high-technology network that links their work, banking, shopping, and communications. That network promises convenience to some; it is also a vehicle for monitoring and controlling or manipulating everyone within its reach.

Meanwhile, those who rely upon high technology to bring new prosperity to the planet, or even to this continent, may soon find their dream shattered. High tech has the potential for destroying more jobs than it creates. Even Silicon Valley, the ultimate boom town of the information era, may eventually self-destruct from the social tension, congestion, and pollution built into its development. In Asia and the United States, workers who assemble high-tech components, the kernels of the "second industrial revolution," labor in conditions reminiscent of the first one.

In the form of high-tech weapons, microelectronics technology threatens the very existence of civilization. Nuclear technology, the recognized instrument of global holocaust, has barely advanced over the past three decades. New electronics technologies, however, have formed the cornerstone of every new strategy for destruction.

■ In January 1985, city officials in Fayetteville, North Carolina, learned from their electronic switchboard records that somebody was sneaking into a city building every night and placing hundreds of calls from two extensions. When the police investigated, reports the *Associated Press,* they found neither burglars nor offending city employees. Instead, they discovered that two computerized Coca-Cola machines, programmed to phone their daily sales totals to a computer at the bottling company offices, were making the calls. The machines, which were supposed to call more than once daily only when they got a busy signal, suffered from what the local Coke service manager called a "manufacturer's flaw."

This story illustrates that if this is the computer age, then it is also surely the age of the technological error. We all seem to find, in the soulless machine, a convenient scapegoat for our human foibles. Yet it is a rare breakdown that can be traced to the interference of a cosmic ray. High-tech machines do only what they have been designed or programmed to do. When they fail, it is the fault of *Homo sapiens.*

The high costs of high tech do not spring from the principles of solid state physics; they are the product of human behavior. Both technology and its uses have been shaped by the social fabric, economic structure, and political imperatives of modern society. Without scientific management, data entry clerks would not be glued to computer terminals eight hours or more each working day. If modern bureaucrats valued civil liberties, the Fourth Amendment would not be obsolete. If a Star Wars laser beam triggers World War III, blame will not lie with a space-borne computer, but with the political leadership that decided it needed to orbit "the button."

During the 1960s when high tech was growing up, Americans treated it with suspicion. Computers were seen as maniacal monsters, like HAL in *2001.* College students carried placards challenging the dehumanization inherent in IBM cards, warning: "I am a person. Do not fold, spindle, or mutilate me!" Even those who learned to program computers were denied direct access to the air-conditioned inner sanctums of corporate, government, and university data-processing centers.

The birth of the microprocessor and microcomputer in the 1970s changed all that. Computers today are inexpensive, accessible, and controllable. They offer shy youngsters the opportunity to communicate and they offer everybody a sense of power. They place, at the fingertips of the computer novice, the power of a 1960s mainframe computer. Even when they are programmed (like R2D2 in *Star Wars*) to fight the wars of the future, they can appear as cute as a puppy dog.

Indeed, microcomputer enthusiasts around the country go so far as to view personal machines as a solution to all the dangers posed to society by the monstrous mainframes. Many hobbyists, programmers, and even the personal computer industry's pioneers were part of the 1960s rebellion against militarism and centralized authority.

Two such pioneers, Steve Jobs and Steve Wozniak, have used their company's resources to portray this image. During the 1984 Superbowl football game, TV viewers saw a startling commercial. The ad opened in a vast hall where scores of zombie-like people sat in rows, listening to a strident speaker whose face was projected onto a viewing screen several stories high. Suddenly a women athlete burst into the hall, pursued by helmeted police. She raced down an aisle and then abruptly stopped. Whirling around, she flung a large hammer at the screen. The hammer sailed through the air, and when it hit the screen, there was a deafening explosion. The picture of the speaker was obliterated. The zombies gasped in amazement as the entire image dissolved. Then an off-camera voice announced that shortly Apple Computer planned to demonstrate why the year 1984 wouldn't be like the *1984* of Orwell's novel.

Apple was announcing the MacIntosh computer, a computer designed, the company claimed, so that anyone could learn to use it in just a few hours. The ad was more than a clever marketing ploy. Apple's top executives actually believe that their machines are a counterweight against dictatorship. In fact, Jobs and Wozniak were so committed to the message that at one point they told their board of directors they would pay to air it out of their own deep pockets if the company chose not to fund the commercial.

The ad also carried a blunt unstated message. It asserted that large mainframe computers had usurped individual control and freedom in the past. In the future, it implied, personal computers will free individuals from institutional control and surveillance.

It is easy to see how individuals who have made their fortunes, or have merely found new autonomy in the world of personal computing, consider the latest in high technology to be the bright side of the chip. As the cost of data processing falls, more people have access to the technology in their homes, classrooms, and workplaces.

However, personal computers, no matter how powerful, are merely tools. Like a hammer, they can build or they can destroy. Although they are tools with unprecedented power and flexibility, to imbue them with the magical ability of producing a necessarily brighter future is simply looking at the world through rose-colored glasses.

The United States is the undisputed global leader in high technology, largely because our society encourages individual creativity. America's competitors, with no comparable source of innovative energy, have not kept up. The big Japanese electronics firms can mass-market high-tech commodities like computer printers and memory chips, yet they have not found a formula that allows them to match their American counterparts in the design of computer hardware or software. Japan trains as many engineers as the United States, but Japanese culture discourages individual effort. Similarly, the USSR expends vast resources to import, copy, and develop high technology, for both military and civilian purposes. But Soviet leaders are unwilling to encourage creativity by permitting the spread of personal computers. Any technology that encourages individualism and independence, they know, will undermine their authority.

In fact, the dependence of high technology upon independent thinkers is society's best hope for reversing the high costs of high tech. Unlike other epochal technologies—nuclear power, for instance—modern microelectronics will never be the sole property of large institutions such as IBM, AT&T, the FBI, and the Pentagon.

Some of America's brightest computer scientists are saying that the Pentagon should not control artificial intelligence research. Technology-wise teachers are developing better curricula and software for presenting computers in the schools. Activists are designing electronic bulletin boards that will bring computerized communications to everybody. This is not to say that high technology will become more humane simply because creative people are constantly helping to shape it. A handful of innovators have already planted the seeds of an alternative vision of a high-tech society. Will we heed their warnings, answer their questions, and support their experiments?

Despite hype such as Apple's *1984* ad, these bright spots on the high-tech horizon are few and far between. The dark side of the chip is winning, for few people recognize the many ways in which new technologies threaten the future of the human community.

Ulysses protected his ship from the Sirens by stuffing his sailors' ears with wax. But the problems of high tech will not disappear if we just close our ears or our eyes. We must, as a society, understand the full impact of high technology if we are to shape it into a tool for enhancing human comfort, freedom, and peace.

Index

INDEX